DOUBLE DEAL

ALSO BY SAM GIANCANA

*Double Cross: The Explosive Inside Story
of the Mobster Who Controlled America*

DOUBLE DEAL

THE INSIDE STORY OF MURDER,

UNBRIDLED CORRUPTION,

AND THE COP WHO WAS A MOBSTER

MICHAEL CORBITT WITH SAM GIANCANA

Wm

WILLIAM MORROW
An Imprint of HarperCollinsPublishers

HarperCollins books may be purchased for educational, business, or sales promotional use. For information please write: Special Markets Department, HarperCollins Publishers Inc., 10 East 53rd Street, New York, NY 10022.

FIRST EDITION

Printed on acid-free paper

Library of Congress Cataloging-in-Publication Data

Corbitt, Michael.
Double deal : the inside story of murder, unbridled corruption, and the cop who was a mobster / by Michael Corbitt with Sam Giancana.
p. cm
ISBN 0-06-019585-1
1. Corbitt, Michael. 2. Criminals—Illinois—Chicago—Biography.
3. Organized crime—Illinois—Chicago. 4. Police corruption—
Illinois—Chicago. 5. Mafia—United States. I. Giancana, Sam, 1954–
II. Title.

HV6248.C6673 2003
364.1'06'0977311—dc21   2002035756

03 04 05 06 07 WBC/QW 10 9 8 7 6 5 4 3 2 1

FOR JOEY

Joey David Corbitt was born on May 25, 1983, and
four short years later I entered a U.S. prison.

Joey was born with Down's syndrome and has
always had to fight for his life. Today he's
eighteen years old and has defied the odds.

Joey is my hero!

FOREWORD

by Kevin Marsh

The people of the United States have a legitimate expectation that the police will fairly and effectively enforce the law, that prosecutors will prosecute without bias and solely upon the evidence, and that judges will administer the law properly and without prejudice. Unfortunately, this legitimate expectation is, in many cases, the epitome of naïveté when it comes to dealing with "organized crime."

Double Deal is a microcosm of a dilemma that exists nationwide. It follows the criminal path of Michael Corbitt, a "street guy" from Chicago who becomes an associate of the Chicago mob. From there, his mob contacts convince him to become a Chicago police officer; and thereafter, the chief of police in a suburban Chicago town. While sworn to uphold the law, Corbitt's real role was to support the goals of his benefactors in the mob. Early on in his life, Corbitt realized that the name of the game was making money—and there was a lot of money to be made in law enforcement by looking the other way regarding crimes committed by his mob associates, tipping off the mob to the activities of Federal law enforcement, obtaining a piece of the proceeds of crimes committed by his partners, and selling "Special Police Officer" badges to known felons, which permitted them to carry firearms.

The mob is all about money. While threats and acts of violence are an integral part of mob business, it is far cheaper to "buy" witnesses, policemen, prosecutors, judges, and politicians than it is to threaten them or do them harm. "Blood is expensive" is a fact of mob life. In the inimitable words of Congressman Kelly in the Abscam investigation, "money talks, and bullshit walks." By paying instead of killing, the mob makes a permanent partner of the person who is bought. Once that person accepts money from the mob, he is owned by them. He must continue to do their bidding or face violent retribution, public exposure of his corrupt deeds, or other punishment for failure to perform his mob function.

The man who said "crime doesn't pay" was clearly not associated with the mob. In point of fact, mob crime pays, and pays well. From the time of Charles "Lucky" Luciano to the heyday of the mob in the 1970s and 1980s, the mob concentrated on giving people what they wanted. From Prohibition booze to gambling, prostitution, pornography, narcotics, infiltration of legitimate businesses and unions, and illicit payments to politicians to buy elections, the mob met the demands of those in need of those services. Even the infamous and often violent area of loan-sharking filled the needs of persons who could not obtain financing through regular banking circles. The mob provided these services to all who sought them, and by doing so generated huge profits for itself. Indeed, during the 1950s, the legendary mob financier Meyer Lansky boasted that the mob was "bigger than U.S. Steel" in terms of profits.

Since all politics centers upon the need for money to use in elections, the mob became adept at buying politicians for their own ends. From the local alderman or commissioner to the most infamous and successful alleged conspiracy between Chicago mob boss Sam Giancana and Joseph P. Kennedy to "steal" the 1960 Presidential election, the mob became masters at purchasing those politicians who were willing to "play ball." The implication of these purchases are ominous, indeed. Politics and politicians are intimately involved in the selection and approval of federal judges, United States attorneys, and directors of the FBI, CIA, DEA, and other law enforcement agencies. The purchase of a "rising star" in the political world could have a disastrous effect upon the investigation and prosecution of organized crime cases.

Michael Corbitt, from an early age, recognized the wisdom of Congressman Kelly's statement about money. The money was easy, provided in cash, and entirely tax free. He coveted the money and the power that it brought, and he willingly and enthusiastically embraced the life of a mobster while wearing the uniform of a man sworn to uphold the law. Corbitt generated substantial sums of illicit money for himself by virtue of his position as a law enforcement officer associated with the mob. During the many years that this association was extant, Corbitt was never prosecuted or even arrested. His downfall came only when his mob associates asked him to commit murder.

Double Deal is a work that should not only be read, but taken to heart for its implications. Michael Corbitt is not the first, nor will he be the last, law enforcement officer to be corrupted by the lure of money and power and to be co-opted by members of organized crime.

INTRODUCTION

The history of America is littered with good guys and bad guys. Like it or not, the red, white, and blue of Old Glory were woven with the twin threads of commerce and corruption. As Chicago mob boss Sam Giancana once said, "One hand washes the other...both hands wash the face."

America's early nineteenth-century cities relied on gangs to maintain the status quo in politics and business, as well as to control prostitution and gambling. Clever entrepreneurs realized that in a civilized society, vicious men required an outlet for their viciousness. In Chicago, this outlet was provided by men like William Randolph Hearst and Moe Annenberg, who built their enormous financial empires thanks to gang muscle.

Hiring criminals to keep the peace or persuade a stubborn populace wasn't anything new; as early as the Civil War, gangs were employed by politicians to assure that voters cast their ballots the right way. This gun-for-hire mentality didn't stop with the cities—when settlers headed west, outlaw gangs were called upon whenever there was trouble.

Eventually, a few of the muscles-for-brains gang members got smart and, by the twentieth century, began replacing violence for the sake of violence—and the petty crime that went with it—with extortion. A simple, straightforward crime, extortion was a highly effective means of making money. Gangs approached vulnerable, immigrant shop owners—as well as proprietors of illicit enterprises—and warned them that unless a protection fee was paid to the gang, their businesses would be destroyed.

The extortion racket flourished across America in the early 1900s and, as a form of organized, criminal enterprise, led to the incorporation of hundreds of small-time gambling and prostitution houses into neighborhood "territories," each representing a sizable source of revenue for the reigning gang. As the stakes grew higher and the dollars larger, bloody wars between rival gangs became commonplace, Chicago's infamous St. Valentine's Day Massacre being one notable example.

Ultimately, the extortion rackets and the power struggles they inspired resulted in a drastic reduction in the number of gangs in America's cities, consolidating power into a handful of criminal "syndicates." As a gangland tactic, extortion changed the *scope* of American crime, but not its face. It was the 18th Amendment banning alcohol, passed by Congress in 1919, which was responsible for that accomplishment.

The lasting legacy of Prohibition would be its impact on the nature of America's organized criminal. Controlling the illegal traffic of alcohol called for a level of sophistication and a type of planning and organization never before seen among earlier thugs. It also required gang members to "come out of the closet"—out of the brothels, alleys, and gaming houses—and into the parlors of "law-abiding" citizens and elected officials who, despite the law, continued to imbibe. Selling moonshine was so lucrative that it quickly turned common street toughs into wealthy, international powerbrokers who—if their criminal activities were discreet, as in the case of men like Joe Kennedy—also became accepted members of society.

Chicago's gangland boss, Al Capone, was well suited to the role of alcohol baron. To acquire large quantities of liquor, handle its distribution, and provide payoffs to police and government officials required a unique combination of traits—from cultivating important working relationships with other gangsters around the country to organizational savvy, cooperation, and ruthlessness. In all of these areas, Capone excelled. Indeed, in 1929, when the Great Depression hit, leaving the city of Chicago with a debt of 300 million dollars and thousands of workers unpaid, Capone and his gang thrived, making millions in moonshine and speakeasies while continuing to expand their power and territory.

Hauling "alky" for the gang offered Chicago's ordinary, law-abiding citizens the means to survive the economic crisis. It also thrust honest men into partnerships with hardened criminals, which in many cases continued long after the financial necessities brought about by the Depression had ended. But far worse than converting a handful of upstanding citizens into career criminals was the partnership Capone's syndicate forged with Chicago's police force—employing officers either directly as bootleggers or, as was more often the case, paying them to look the other way. It was this underworld alliance with law enforcement that would dog the Chicago's legal system well into the future.

With the repeal of the 18th Amendment in 1933, the annual take of America's gangs was estimated at a staggering 18 billion dollars, an enormous figure considering that the entire United States budget was only 4 billion. By making something illegal that virtually everyone wanted—from high-class statesmen to low-class sweatshop workers—the U.S. government had managed to transform penny-ante street gangs into a sophisticated and wealthy network of organized criminals with thousands of politicians and police officers on its payroll. In just over a decade, the line between the bad guys and good guys had been blurred forever. Now they were the same guys.

⊕ ⊕ ⊕

By the time young Sam Giancana resumed life among his peers in the Chicago Outfit in 1942, he'd established his reputation as a violent and ambitious gangster. With four years in Leavenworth federal prison for moonshining behind him, he now turned his attention to moving up the organizational ladder.

In Capone's gang, where the amount of money a man earned for the organization was considered the only real measure of his success, Giancana had the inside track: an open door into Chicago's million-dollar-a-day colored policy and numbers rackets, thanks to a jailhouse bargain he'd struck with the city's black policy king, Eddie Jones.

In 1944, flush with cash from the policy rackets, Giancana made his move, launching an aggressive push for territory that would become known as Chicago's bloodiest gang war since the old days of Prohibition. Since his release from Leavenworth, Giancana had assembled an army of lieutenants comprised of boyhood friends and fellow members of the notorious "42 gang," men who were as vicious as he was and more than willing to do whatever he asked. For months, Chicago's streets and alleys were the scene of bombings, murders, kidnappings, and ambushes. Newspaper headlines screamed with sordid tales of gun battles in broad daylight and bullet-ridden bodies found floating in rivers or tortured and stuffed in trunks of abandoned cars. When it was finally over, more than a dozen of Giancana's competitors—including his benefactor, policy king Eddie Jones—had been eliminated from the playing field.

Among officials in law enforcement, Sam Giancana had earned a place among other gangland "public enemies" as a vicious killer and threat to society. But to the men he'd sought to impress—Capone successors Tony Accardo and Paul Ricca—Giancana was a man whose ruthlessness and talent as an earner demanded their respect, a man whose lust for power and money they could not afford to ignore.

In 1946, when New York crime boss Lucky Luciano called a meeting of the Commission—a consortium of the Big Six syndicate families and their thirty-six bosses—and included Sam Giancana on the roster, it was patently clear just how far the young Chicagoan had come since his old "42 gang" days of playing wheelman for Accardo and Ricca. When Ricca was sent away to prison that same year, making Accardo boss of the Chicago Outfit, it was also clear that Giancana was the only logical choice to succeed Accardo as underboss.

Over the next decade, Giancana set up an enormous jukebox, pinball, and vending machine operation comprised of more than twelve thousand machines and five hundred "distributors." With each machine bringing in ten dollars a week—a six-million-dollar annual take—it was a skimmer's paradise. During this period, Giancana also expanded his power base outside Chicago, cementing the Outfit's hold on virtually all of the territory west of the Mississippi with the exception of Las Vegas, which was considered "open" to organized crime.

In his drive for expansion, Giancana developed influential relationships with bosses from other cities, growing especially close to New York's gambling kingpin and political fixer Frank Costello. The two men collaborated on a number of successful ventures, but it was their international collaboration—and Costello's accompanying introductions to a variety of foreign and U.S. officials—that would prove most valuable to the young Chicago mobster, paving the way for the Outfit's multi-billion-dollar overseas gambling empire, an accomplishment that would make Giancana the undisputed successor as boss of the Chicago Outfit.

⊕ ⊕ ⊕

In 1944, while Sam Giancana was waging a war on his gangland competitors in Chicago, the United States had launched a war of its own. And once again, just as they had in the days of Prohibition, the events of history would thrust organized criminals and members of law enforcement into partnership.

Intent on defeating the enemy, U.S. intelligence and the military turned to anyone who could further their cause. Members of the American Mafia offered a special opportunity, given the considerable influence these criminals continued to maintain with cronies in Europe and Asia. Among the better documented examples of these mob confederates and the role they played in aiding the U.S. government was that of New York gangster Lucky Luciano.

Utilizing his considerable network of contacts in organized crime, both in the United States and abroad, Luciano aided the Office of Strategic Services (the intelligence forerunner to the CIA) in a number of covert operations, including the invasion of Sicily by allied forces. Similar relationships with underworld figures were established in the Pacific and the Philippines and, by the close of World War II, organized crime had a number of "friends" in high places.

In 1959, history would repeat itself once more in Cuba, when U.S. intelligence and organized criminals were again faced with a common enemy—in this case, the island's new revolutionary leader, Fidel Castro. Although their covert attempt to oust the young dictator was a failure, it managed to solidify their relationship, leading to a sustained partnership between organized crime, in particular Sam Giancana's Chicago Outfit, and U.S intelligence.

With the advent of the turbulent sixties and the Vietnam war, the Outfit's illegal activities overseas began receiving high-level protection from members of U.S. intelligence. In return, the CIA and its counterparts acquired millions in illicit funds, much of it used to further secret operations overseas.

By the time the Iran-Contra scandal broke in the 1980s and all eyes had turned to Central America and CIA dirty tricks, the Chicago Outfit had become part of the rarefied world of U.S. national security and intelligence, a world where the assassination of world leaders could be justified under the banner of American patriotism, a world where corruption and capitalism worked hand in glove, creating a new global economy dominated by criminals and corporations. And in an organization where money was the measure of a man's success, it was also a world tailor-made for the mob.

⊕ ⊕ ⊕

It's been said that the name of the game in the Chicago Outfit is money. But the fact of the matter is, it's the *only* game. This book is about one man who took a gamble and played that game. "Give me a man who steals a little and I can make money," Sam Giancana used to say. And indeed, there were hundreds of men under Giancana's rule who fit that description, among them a young punk from Chicago's South Side named Michael Corbitt. Like his mentor Sam Giancana, Corbitt had an overriding affection for the greenback, an affection that subjugated both morality and law.

As a product of the fifties, that postwar golden age when the good life was supposed to be available to everyone, Michael Corbitt is the quintessential example of the American Dream come true—even if you have to become a criminal to get there. The consummate double dealer, throughout his life Corbitt has played both ends: becoming a highly decorated Chicagoland police chief, a trusted mob insider, and a hell of a high roller.

As such, Corbitt dishes out a tantalizing firsthand account of organized crime. His story is deeply personal—as well as deeply disturbing. Invited into his innermost psyche, one cannot escape an uncomfortable truth: Michael Corbitt is, in many ways, the American Everyman. He's your next-door neighbor. He's you. And me. If we're honest, we might even admit that Michael Corbitt is the person we might have become if only we'd had the chance. After all, who hasn't wanted that showroom car? The house, the boat, the big bank account? Sounds good, doesn't it? There's just one catch: You'll have to work for the mob. And for almost thirty years, that's exactly what Michael Corbitt did.

For a while it looked like Michael Corbitt had it all. Then his luck turned. Things got ugly—and deadly. In 1989, he was charged with four counts of racketeering, among them conspiracy to commit murder. In the end, he got twenty years in federal prison and, as a bonus, a contract on his life.

Why the contract? Because, quite simply, Corbitt knows too much. From the glory days of Sam Giancana through the high-flying seventies, eighties, and beyond, Michael Corbitt has been witness to it all. And that's where I come in. That's where *Double Deal* begins.

AUTHOR'S NOTE

Whether anybody likes it or not, this is a true story, every god-damned word of it. And if having it come out means I have to jackpot somebody, I don't care, I'm gonna do it. Now, sometimes I think, well, maybe I don't wanna do it. I mean, after all, some big people are gonna go to jail over this book. Then I think, What the hell am I talkin' about? Forget jail . . . there's a good chance somebody's gonna get whacked over this book. But hey, if they do, well . . . fuck 'em. Those are the people I don't give two shits about. That's right. I don't care. Far as I'm concerned, they had it comin'.

—*Michael Corbitt*
August 2002

CHAPTER 1

He's gotta go."

The words on the tape player ricocheted off the prison walls. I knew that voice like it was my own brother. It was Sal Bastone, the big greasy-headed son of a bitch who ran the Chicago mob's North Side crew, the guy the FBI called a high-ranking Outfit "loan shark and racketeer." Sal and I went way back—to the sixties, when I was on the police force in Willow Springs, Illinois.

It may be hard to imagine a cop being involved with the Outfit, but a guy in that position can be very useful to organized crime, and I was no exception. Like a lot of underpaid police officers, I was more than willing to "look the other way." But during the ten years I worked with the Bastone crew, I became more than just a crooked cop who turned his head for a price. I became Sal Bastone's bagman, his gun man, his driver, and his courier—as well as his best friend.

The tape player was still on, but it wasn't Sal Bastone I was hearing now. Another voice said, "Gotta go? You mean . . . GO?" It was Gerry Scarpelli and he sounded shocked. "You sure we're talkin' about the same fuckin' guy? I mean, Jesus . . . this is your fuckin' man out west we're talkin' about, right? Your goddamned driver . . . MIKE?"

My stomach turned. I couldn't believe my ears. Sal was telling Gerry to whack me. *Me.* And I was supposed to be Sal's best friend. At least I'd thought I was. Of course, unlike Gerry, I wasn't about to reveal my feelings on the matter—not with two FBI agents sitting right across the table from me, watching my every move.

"Yeah," Sal said impatiently. "*Yeah . . . we're talkin' about the same guy . . . MIKE. Like I said, he's gotta go. We got a major problem here. I don't know if you seen he got indicted. But some people are concerned about the situation . . . you hear what I'm sayin'? So that's that. He's gotta go.*"

Gerry's voice went shrill, "*Okay, okay . . . I hear you. But you know, his goddamned kid is with him wherever he fuckin' goes . . . what about the kid?*"

At the mention of my son Joey, my stomach did another nose-dive. Sal Bastone could have me whacked, fine. But a defenseless little kid? Joey has Down's syndrome. He's just a baby. He's also Sal Bastone's godson. I couldn't imagine that Sal would ever go along with anything that might hurt Joey.

Sal went on, "*Hey, I don't give a rat's ass who the fuck's with him. Like I said . . . the cocksucker's gotta go. Okay? You fuckin' got that? So do what you gotta do. Just get the job done. This isn't just me talkin'. . . . This is comin' from up top, from you-know-who . . . you understand what I'm sayin' here?*"

I guess having the order to take me out come from Chicago's top man on the street, Joey Lombardo—who Sal had always referred to as "you know who"—lessened the shock to some extent. I wasn't surprised to hear that Lombardo wanted me taken out; he'd always hated me. Forget any loyalty I'd shown to the Outfit, I'd always been a badge to him. And I wasn't surprised that Sal would go along with the idea, either. In my line of work you have to be a realist. Business is business. So I could live with Sal's betrayal of our friendship. It was his total lack of compassion for my son Joey that I couldn't handle.

One of the agents hit STOP on the tape player. "So there you have it," he said. "What do you think of your old friend Sal Bastone now?"

Since 1989 the Justice Department had tried everything they could think of to get me to turn and go into the federal witness protection program. Up to this point, nothing had worked. But I guess they figured that hearing my best friend planning to whack me and my kid would do the trick, that Michael Jerome Corbitt, Tallahassee inmate 96751-024 and former Willow Springs chief of police, would finally cooperate, go WITSEC, and disappear into Numb Nuts, Iowa.

The other agent shook his head and sighed. "Sounds like Sal Bastone doesn't care who's in the way when they come after you, Mike . . . even if it's your kid."

I glared back at him.

He threw up his hands and shrugged. "Hey, man, don't blame us. . . . It wasn't our idea to play the tape. You wanted to hear it."

His partner nodded. "And now that you've heard it, what are you gonna do about it?" He leaned forward and folded his hands on the table. "You tell me, Mike . . . what's it gonna be?"

Our eyes met and then—*whoa*—it was déjà vu. All of a sudden, no more agents. Instead it was Sam Giancana, the infamous Chicago mob boss, staring back at me. "So what's it gonna be, kid?" Giancana asked. His dark eyes were shining out from under a black fedora. "You wanna be a cop?"

A chill ran up my spine. It had been years since Sam Giancana asked me that. Back then, in the sixties, I was just an overgrown juvenile delinquent with a leather jacket and a lot of attitude. I'd looked up to Outfit guys like they were gods. So of course I'd said yes to Sam Giancana—even though I didn't have the slightest desire to be a goddamned cop.

"Good," Giancana said. "But just remember, kid . . . don't forget who your friends are." And then Sam Giancana winked, like me and the world's greatest mob boss shared some big fucking secret.

It was one of those moments in life when you suddenly realize that from then on, nothing is ever going to be the same again. *A moment just like this one*—

The agents across the table came back into focus. It was true—nothing would ever be the same again, not after hearing that tape.

In five minutes, everything had changed, including the rules. What was it Sam Giancana had said? Don't forget who your friends are? What friends?

One of the agents interrupted my thoughts, repeating his question. "So what's it gonna be, Mike? You'll have the Bureau's full protection."

"I'll tell you what it's gonna be," I yelled, beating my fist on the table. "Fuck you. And fuck your goddamned FBI protection, too."

The agents both smiled sympathetically. "But you care about your family, don't you?" one said. "You care about your boy?"

My boy? Well, that did it. They weren't going to use my little Joey as a pawn in their game. I'd heard enough of their bullshit. I stood up, slamming my chair against the table. "We're through," I said and headed for the door.

⊕ ⊕ ⊕

That night in my cell, I tried to shake the image of the agents' faces when the prison guard escorted me out of the conference room. They thought I was a dead man. Hell, maybe I was.

That was one of the longest nights of my life. For most of it, I sat on my cot, shaking like a frigging rabbit. It wasn't the idea of getting whacked that made my hands tremble so badly that I couldn't hold a cigarette. I wasn't afraid of some mob hit man. It was blind rage, that's what it was.

Hearing my "best friend" Sal Bastone put a hit on me had been bad enough, but the fact that he didn't care if my little boy got killed in the process put me over the edge. If I was afraid of anything, it was that something might happen to my son. I stood up and took a deep breath, catching a glimpse of my reflection in the cell's steel mirror. And then Sam Giancana's words came echoing back through time. I could almost hear him saying, "What's it gonna be, kid?"

The face in the mirror was staring back, waiting for an answer. Yeah, what *was* it going to be?

I was going after Sal Bastone and his entire crew, that's what. I'd make them pay. But how could I get to them? The truth is, I'm no killer. I might've gotten nailed for conspiracy to commit murder, among other things, but that was a complete fabrication, made up by a bunch of ambitious attorneys and the media. Sure, I took down plenty of men in my time, but it was always in the line of police duty. And it wasn't like I *wanted* to kill them, either. I'd had to do it.

But this situation with Sal Bastone was totally different; he deserved to die. And so did the guy above him—the man who'd taken over after Sam Giancana's murder and continued the Outfit's collaboration with foreign governments and the CIA. In the sixties, the McClellan Committee had called him "a mystery man." And he still was. But he wasn't going to be much longer. I wasn't sure how, but I was going to find some way to blow the lid off his world. And I didn't care if I died doing it, either.

So that was it; I had my answer. That's what it was going to be. I looked back at the reflection in the mirror. And then I winked.

CHAPTER 2

At heart my parents were just good honest farm people who went to church whenever the door opened. In some ways they were practically saints. But they weren't soft; they were strong-willed and hard as nails. In 1939, during the Great Depression, they lost their first child, Luella, to pneumonia and hard times. Even then they didn't give up trying to make a go of it on the family farm down in southern Illinois, but by 1942, when things still weren't looking up, Dad loaded the family into the old pickup and headed north.

When they got to Chicago, they rented a little converted garage in the blue-collar town of Summit. Dad got a good factory job and made a little extra on the side doing carpentry work around the neighborhood. By that time there was a world war going on, but even so, their future was bright.

They already had two little girls, Gina and Terry, so when Mom got pregnant that year, Dad prayed for a son. And on March 17, 1944—Saint Patrick's Day—his prayers were answered. They named me Michael Jerome.

⊕ ⊕ ⊕

Given my parents' backgrounds, it's no surprise that they were hard on us kids. Mom and Dad were very harsh disciplinarians who also had hair-trigger tempers. By the time I was seven, in 1951, I knew that if I whined or talked back, I'd get a backhand or the belt. It might have been the fifties, but I wasn't living with Lucy and Ricky.

By that time I also had two more sisters, Cecilia and Ellen. I was the only boy in the family at the time—my little brother, Tim, wasn't born until 1964—and my sisters roughed me up all the time. I hated it. I wanted to beat the crap out of all of them. But that was another thing I learned early in life: beating up on the girls wasn't an option. If I did that, I'd get it from my parents, too.

Clearly, the Corbitt household wasn't the kind of environment that made a kid want to express himself. By the time I entered grade school at Saint Blaise, I was one shy, insecure little boy. For the first three years I got along by sitting in the back of the classroom with my nose stuck in a book. But in 1953, when my parents announced they were sending me to Graves Elementary because we couldn't afford parochial school, my attempt to stay "low profile" didn't work anymore. For one thing, I was big for my age. For another, without those Catholic school uniforms to hide behind, it was obvious just how poor my family was. In fact, we were so poor that to make do my mother dressed me in my sisters' hand-me-down blouses—which made me the target of every bully within a fifty-mile radius. Pretty soon, black eyes and bloody noses were a way of life. But I didn't get any sympathy at home. My sisters laughed at me, and my parents ignored my plight entirely.

It was a very lonely time, and like a lot of unhappy kids I withdrew to a world of make believe. Mostly I spent my time in front of our black-and-white TV watching the cowboy shows. I guess it was only natural that a kid who felt so weak and vulnerable would be drawn to westerns, with their larger-than-life heroes. My favorite was *The Lone Ranger,* with his white horse and black mask and six-shooters. He always got the bad guys. I used to imagine that someday I was going to be just like the Lone Ranger. Yeah, *someday,* I was going to be a hero, too.

⊕ ⊕ ⊕

I can still hear "Rudolph the Red-Nosed Reindeer" playing on the radio that fall afternoon when we kids sat on the living room floor and made up our Christmas lists. My list was usually a mile long, but this time I wrote down one thing: Lone Ranger cap guns and a belt. I figured that with those little beauties slung over my hips I'd be just like my TV hero. Of course, I didn't have any hope of getting them. Dad always made our Christmas presents. But he couldn't make what I wanted that year with a hammer and nails. And there wasn't any chance he'd buy it, either. Other kids like Donnie Carter down the block might have all sorts of toys, but me? Like Mom always said, money didn't grow on trees.

You can imagine how surprised I was when I woke up on Christmas morning and found beautiful pearl-handled pistols and a black leather holster under the tree. I spent hours in front of my bedroom mirror, twirling and drawing those guns. I even slept with the damn things. I was the happiest kid in the world for the next two weeks. But then, one wintry afternoon, when I was out in the front yard practicing my quick draw, the boom fell. And as luck would have it, it was Donnie Carter who came marching down the sidewalk and ruined it all.

"Those are my guns," he declared, balling up his fists.

I held my ground. "No they're not. I got them for Christmas."

Donnie got a good laugh out of that. "You are sooo stupid," he sneered. "So you got 'em for Christmas . . . so what? You mean you can't tell they're used?"

I looked down at the pistol in my hand. They weren't new. I saw that now.

At the look on my face, Donnie grinned. "That's right, dummy. Your father bought 'em from my dad. Gee, Corbitt, you *are* a dope."

It was like I'd been hit in the gut. Dad hadn't bought the belt and guns from a toy store after all. They'd been Donnie's to begin with—which to my way of thinking meant they weren't *really* mine.

I tried not to cry when I gave them back to Donnie. I didn't want another kid's hand-me-down toys—I wanted my own.

From then on, I had no use for Christmas. When we made out our Christmas lists, I left my paper blank. They say that by the time a kid is six or so, his life is set. And looking back, I think that's true. Since the day I gave those cap guns back to Donnie Carter, money has been the focus of my life. You could even say it's been my demise.

CHAPTER 3

By 1955 the Lone Ranger was just a childhood memory. James Dean was my idol now. After *Rebel Without a Cause*, all the girls, including my sisters, went crazy over him. So naturally there wasn't a red-blooded American male who didn't try to look like James Dean. I was only eleven, but I rolled up the sleeves of my white T-shirt. I slicked back my hair. I even tried to smoke. But it wasn't just Dean's style that attracted me. I identified with his angry young man image. And why not? After all, I was one angry kid—mostly about my family's financial circumstances. Coincidentally, it was at this same time that my frustration with not having the same material things as the other kids finally got to me.

I knew that if you didn't have money, there were other ways to get the things you wanted—so I became a thief. And I got very good at it, too. It was the start of a great career. Someday I'd be made-to-order for the Outfit, but back then, someday was a long way off, and, except for the thrill I got from my adolescent crime sprees at the corner store, my life went on pretty much the same. For the next two years I went to Graves Elementary, and if it was possible for a kid to get used to feeling like a dope, I did.

In 1957 I was thirteen and practically invisible. It didn't look like there was any way an overgrown kid with an inferiority complex to match would ever have his day. But then fate intervened and changed everything.

⊕ ⊕ ⊕

The factory where Dad worked—Visking—had an annual picnic that we Corbitts looked forward to. Every year the employees and their families—there were over a thousand in all—got together on a lake in a nearby forest preserve. It was a kid's dream come true: there were hot dogs, hamburgers, pies, cakes, and homemade ice cream as well as all types of games and entertainment. We rode ponies, played baseball, went on hayrides. The Visking picnic was always good old-fashioned fun. But on that particular afternoon in 1957, the fun almost turned to tragedy.

I happened to be standing at home plate waiting for the next pitch when somebody yelled that one of the hay wagons had flipped over and thrown all the little kids into the lake. Hearing their screams, I dropped my bat and started running. My baby sister Cecilia was on that wagon.

It was a deep lake, and I figured they'd all drown if somebody didn't get to them right away. Before I knew it, I was up to my neck in water. I was so determined to save my baby sister that I completely forgot I couldn't swim. Somehow I made it to every one of those little kids and got them back to shore, where a big crowd had gathered. It was quite a scene. All the mothers cried and hugged me. The fathers shook my hand. Everybody called me a hero.

For days after the picnic, I walked around on cloud nine. There was even a story about the incident in the local paper. Dad puffed up like a rooster. It was like I was somebody else, a new Mike Corbitt. But when I walked the halls of Graves Elementary I felt more like the Platters' "Great Pretender" than a hero. I knew the truth: Michael Corbitt wasn't brave at all. He was a coward and a big clumsy kid who had no use for fighting and hated the sight of blood even more—especially his own.

To make matters worse, my new celebrity only brought my tormentors out of the woodwork; getting beat up after school became a daily event. Sometimes I'd get lucky and just catch a lick or two.

But now and then things got brutal—like the time four big Polack boys beat the crap out of me. By the time I got home from school that day, I was really hurting. What I needed was a shoulder to cry on, but all I got was a lecture from my oldest sister, Terry. Terry had no patience for a snot-nosed, whining kid like me. "Don't you ever come home like this again," she shouted over my sobs. "You do and I swear I'll give you a beating myself."

I didn't know if I was more hurt or mad. But I knew one thing for sure: I hated her. And my family, my school—yeah, I hated them all. Alone in my room, I cried for hours. When I finally stopped crying, the anger set in. That's when I promised myself that from that day on, if anybody tried to mess with me they were in big trouble.

As it turned out, I got to make good on that promise the very next day when a kid from my neighborhood, Roger Douglas, started busting my chops out on the playground. Roger was a real bruiser, and I'd always tried to stay out of his way. But that day I was in no mood to dodge anybody. I can't remember what he said, but it didn't matter. All that pent-up rage from all those years of beatings came pouring out. I went for Roger Douglas with everything I had.

All the kids gathered around egging us on, but I didn't need any encouragement; I got Roger on the ground and just kept hitting him. I was loving it, too. When it looked like I might actually win the fight, five of his friends jumped in and tried to pull me off. But there wasn't a chance that was going to happen; I turned on those kids like a mad dog. We exchanged some good blows, but I didn't even care that I was hurting. I kept right on hammering those five guys, and when I was done with them, I went back to Roger.

I gave it to Roger real good. He was spitting blood and teeth all over the place. Finally he yelled, "Stop." And that was that. I just got up and walked away. All the kids were yelling at me to come back, but I just kept walking. I didn't even bother to look back.

⊕ ⊕ ⊕

I figured the fight with Roger Douglas would get me in hot water, that I'd get a beating from Dad and maybe even get expelled from

Graves. But instead I was the object of praise. "You gave it to 'em pretty good, huh, Mikey?" my sister Terry crowed at the dinner table. Dad nodded. "Sounds like those bullies had it coming."

It was the same at school; the principal didn't say a word. And when I walked down the halls, it was like I was Moses at the Red Sea, parting the waters. Everybody stepped aside when they saw me coming. Thanks to that fight, I went from being a zero to number one, big man on campus. Roger Douglas, the guy I beat up, became my best friend in the whole world. And so did his buddies. I started hanging out at all the coolest joints—places like Lucky's diner and Central Billiards where the gangs stood out in the parking lots looking for trouble. I began smoking cigarettes on the sly. I even made it onto the school basketball team.

I started to change in ways I couldn't have imagined. I dropped my little thefts at the corner store and started going for bigger stuff. Kids gave me lists of the latest hit records they wanted me to "pick up" for them at the store. I was swiping bottles of perfume right off department store counters. In drugstores, I cased the joint and then stuffed a few packs of cigarettes, some candy, and a half-dozen *Life* and *Photoplay* magazines into an oversized parka.

Like all criminals, the more successful I was, the bolder I got. I started walking into clothing stores and taking shirts, jeans—whatever I wanted—right off the rack. Sometimes I felt bad about it, but not that bad. After all, my new career was making me the most popular guy around. Those were some great times. Now I was somebody. I even learned to dance and got so good that I started winning contests. Pretty soon, girls were asking me out.

It's a wonder my head didn't explode. But even though I was on top of the world, I knew there was something wrong. Deep down I still felt like the Great Pretender. Sure, on the outside I had all the right moves. But inside I was afraid—afraid they'd find out who Mike Corbitt really was: a sad, angry kid who had to steal to get the things they only had to ask for.

⊕　⊕　⊕

The city of Chicago and the family farm were like two different planets. Being in Chicago was a rush—even as a thirteen-year-old kid, I got a charge out of the streets. I loved the noise, the fast pace. And I loved the street-corner brawls that guys back then called "rumbles." On the farm, things were real slow. Forget about rush. If a pretty girl smiled at you, that smile stayed with you for a long time—not at all like in Chicago.

As you can imagine, the time we kids spent on the family farm every summer was a real whack in the head for us. Dad called it our "working vacation." I learned to drive a tractor. I bailed hay, did some cultivating. Talk about muscles. After a few summers down south, I had muscles in places most city kids don't even know exist. I was strong, too.

Of course, it wasn't all work at the farm. We had a big time fishing and swimming and riding horses. After a while I'd even forget the words to "Wake Up Little Suzie" and the steps to the Stroll. In a matter of days down on the farm, a guy turned into a total square. And the funny thing was, you didn't even mind.

Occasionally our family went to the farm for Christmas. Those were great times. I loved it down there at that time of year. The way the snow stayed clean, in drifts as white as sheets—nothing like in the city, where the snow turned into a big pile of black mush. I loved the smell of the country, too. There weren't any factory fumes hanging over your head. The air smelled like the Christmas trees we cut down each year.

It was weird the way material things didn't count for much on the farm. I even left my trusty parka back in Chicago. Somehow I didn't feel so poor in the middle of nowhere. And when it came to Christmas presents—they didn't matter, either. I hadn't asked for anything for Christmas since my old Lone Ranger days.

But on Christmas morning in 1957, I realized that although I could pretend I didn't care about gifts and the like, deep down—well, I was still a sucker for a present. And that year, when I saw that funny-shaped package under the tree, my heart skipped a beat.

"Go on, Mike, open it up," Dad said. He smiled and leaned forward in his easy chair.

When I saw that sixteen-gauge shotgun, I let out a war whoop you could've heard all the way to Chicago. It was just what I wanted. Even though it was snowing like mad, Dad and I headed for the fields. It was my first hunt, and I started wondering how it was going to feel to actually kill something. I didn't have long to wonder. All of a sudden Dad yelled, "Fire," and I did, and *bam*, I'd killed a rabbit. I didn't feel any remorse. No guilt. No sympathy. Nothing. It was alive one minute and dead the next. And I was the reason. It was as simple as that. Case closed.

But that wasn't the end of it for Dad. I'd never seen him so elated. "You're a man now," he said and shook my hand. Looking back, I see what a special day that was; we were never that close again.

A few days later, we packed up the car—an old gray Dodge that Dad loved and I hated—and got ready to head back to Chicago. The snow was coming down pretty hard by the time we all piled in and Dad turned the key in the ignition. Naturally, the engine didn't turn over. No surprise there. The car was a lemon. But for some unknown reason, Dad couldn't see it, which just drove me nuts.

Dad looked at Mom, gave a shrug and shook his head, and climbed out of the car. He was just lifting the hood when Mom told me to go lend him a hand. I stood out there forever, shivering in the cold, watching Dad screw around with the goddamned distributor. He was an embarrassment. The last thing on earth I wanted to be was a guy like Joe Corbitt. All you had to do was look at the way he lived. It was all get up and go to work, punch in, punch out, come home, go to bed. Then start the whole thing all over again. That was no life.

It wasn't just being a factory worker that sucked, either. I figured it wouldn't have made any difference if Dad had stayed on the farm. Things would've been the same—only on a farm he would have been nursing some piece of dirt that never gave him a thing back.

That's when I realized I wasn't going to find what I wanted on a farm or in a factory. But I also realized something else: Good things didn't just come to you, you made them happen. Yeah, with force you could take what you wanted. And if that didn't work, you could steal the rest.

The old Dodge engine turned over, and Dad closed the hood. We looked at each other, but we didn't say anything. There was nothing to say. We climbed in the car and headed for home.

CHAPTER 4

During the winter of 1958 the McClellan Committee was getting in full swing, and like everybody else in America, Dad was glued to the TV and the committee hearings they were recapping on Channel 5. I was just a kid at the time, so none of it really mattered to me. But when they started talking about Chicago and its "colorful gangster underworld," well, then I got interested.

All of a sudden guys like Al "Scarface" Capone, Paul "The Waiter" Ricca, and Sam "Mooney" Giancana—guys Dad called "common thugs"—were practically legends. To a kid, it was very strange. And it was confusing, too. With all the attention those guys were getting on TV, you had to wonder if they were all that bad. And I wasn't the only one confused about who the bad guys were; that year, even songs on the radio idolized outlaws. "Tom Dooley" was a big hit, and it glorified a killer.

Every day there were stories in the paper and on TV about how powerful the Chicago mob was. Mostly it was about how much money their rackets were making. They said it was in the millions. To a kid who wanted money more than just about anything, that was really impressive. Another thing that impressed me about the mob guys I saw on TV was that they hadn't been rich to begin with. They'd all grown up poor, like me. In some strange way, they were actually inspiring.

Of course, I understood that the men in dark glasses who I saw testifying before the McClellan Committee were hardened criminals. I'd heard the stories—the ones about mobsters and contract killers and hit men. But they didn't look that scary. And from what I could tell, they weren't suffering too much for their crimes, either. They weren't behind bars. It looked like crime paid. And paid well.

Maybe it was seeing all those rich mob guys on TV, but in any case, my resentment toward Dad had been growing. Things were very tense between us. I blamed him for our poverty. I thought he was a sap, and because of that I gave him no respect whatsoever. The fact that I was ignoring "house rules" and staying out late drinking didn't make it any better.

Things finally came to a head one night. It was late, and Dad was waiting up for me. The minute I walked in the door, he started in with a lecture. It was always the same: I was only fourteen. Drinking was against the law. It was a school night. Blah, blah, blah. Through the whole deal, I barely looked at him. And when he was done, I shrugged and rolled my eyes and gave him my best "fuck you" look.

That did it; his face turned purple and he jumped out of his chair. "You're grounded," he shouted.

"Grounded?" I said with a laugh. "I don't think so."

At that remark, things went to a whole new level. All of a sudden Dad and I were toe to toe. Up until that moment I hadn't realized how much taller I was than Dad. My father was looking *up* at me. The TV was crackling with that old off-the-air sound they used to make. Otherwise it was dead quiet. We stood there, at an impasse, not saying a word, for what felt like an eternity, until finally Dad sighed and took a step back. "Go on . . . go on to bed," he said weakly and waved toward the hall.

It was the first time my father had backed down. From then on, I wasn't afraid of his belt. I realized my folks couldn't *make* me do anything. Overnight, I was my own man. Down the road, my folks would make one last play to get me to tow the line. But for the time being, it was all a bunch of idle threats. Sometimes I think they just gave up. But other times I think that maybe, deep down, they realized that it was just too late.

⊕ ⊕ ⊕

In 1958, where I came from in Chicago, being in a gang was a big deal. My gang was made up of neighborhood kids—guys like Roger Douglas and our buddies from school. Mostly it was Italians—hot-tempered greasers who wore leather jackets and white T-shirts with pegged jeans and pointed-toe boots. Naturally, I dressed the part. I spent hours in front of the bathroom mirror, making that one curl fall just right over my eye. When I was done, my hair was so loaded down with Brylcreem that nobody could ever accuse me of listening to that TV jingle that said "a little dab'll do ya." A *dab* of Brylcreem? They weren't going to call Mikey Corbitt a greaser for nothing.

I never figured out why it was mostly the Italian guys who went for the gang thing. Maybe they'd all seen *The Wild One* a few too many times. Of course, not all the guys were Wops. There were other ethnic groups. There were a couple of Anglos, a few Greeks, some Polacks to round things out. Nobody cared that I was Irish and not Italian. And it didn't matter to me. It felt right, hanging with the Italians. As far as I was concerned, I *was* Italian. Being in the gang meant I had a new ethnic heritage and a new family, even if it was all made up. And I had a sort of home away from home at all those pool halls and dives where we hung out. School and that other family—the real one—were minor annoyances.

The guys in my gang weren't criminals, they just acted tough. We never toted guns or held up a joint. Sure, we'd been known to hot-wire a car or two, which was one of Roger's specialties, but it would be several years before any of us graduated to the majors. A few of the guys, among them Roger, would eventually go for the real deal—stickups, car heists, and burglaries. But back then, that was way out of our league. We were playing at being gangbangers. What mattered to us was the petty thefts and back-alley rumbles. There was always somebody's ass we could kick. And kicking ass— well, that was *my* specialty.

CHAPTER 5

The first time I laid eyes on Pete Altieri, I knew he was one of them. Screw James Dean and Marlon Brando. This guy was for real. He might not be on the ten o'clock news, but I was pretty sure that Pete Altieri was in the syndicate.

Pete owned Lucky's, the local diner and teen hangout, at 63rd and Harlem. Across the street, he had A&W Electric, an appliance repair shop—at least that's what the sign said. Occasionally I saw guys unloading a couple of refrigerators and a stove or two out front. Meanwhile, in back it was like Grand Central Station: there were trucks and vans coming and going night and day, unloading all sorts of coin machines and games. And then there were the guys who hung around the place—they weren't exactly your Maytag man. These guys would walk in the back door empty-handed and a few minutes later come out whistling, carrying canvas bags the size of loaves of Wonder Bread.

Even though I figured A&W had to be a front for the syndicate, I couldn't walk up to Pete Altieri and ask, "Hey, man . . . you in the fuckin' mob?" I wasn't stupid. Besides, I hadn't seen anything going down at A&W that was actually against the law. So what if they handled appliances in the front and coin-operated vending machines in the back? I hadn't seen a single slot or pinball machine, which at that time *were* illegal.

I couldn't shake my feeling about the place, but it was still all in my gut, until late one night in June when a couple of big semi—tractor trailers pulled into A&W's back lot. Next thing you know, that little repair shop was in high gear. It was like Pete Altieri was running a top-secret military installation. There were guys all over the place, hauling ass. Before the night was over, they'd unloaded dozens of pinball and slot machines. Talk about hitting pay dirt. I'd hit the mother lode. Forget about having a feeling about the joint. Now I had the facts.

There were a few other facts about the machine business that I found intriguing. According to the *Chicago Tribune*, the syndicate's coin-operated machine racket had been completely overlooked by law enforcement. The paper said the mob was taking in at least four hundred thousand bucks a week from the twenty thousand machines they had scattered around the state. I'd find out later that they were bringing in more like two million a week. That's a hundred and four million a year. It sounds like a lot, like maybe it's an exaggeration, but it's not. You have to consider that everything from cigarettes, candy, and colas to condoms and games are sold out of coin-operated machines. And forget about all the illegal machines; add up the legit business, and you can see how much money is out there. There are millions to be made in vending. And like I found out later, those Chicago boys squeezed out every last dime.

It looked like Pete Altieri and his pals were onto something big all right. And I wanted in on it, too. But how could a punk kid like me ever get next to those guys? Talk about a long shot. But then again, every country boy on a hunt knows that all you've got to do to bag your prey is get close. And talk about close—I was right across the street, at Lucky's.

⊕ ⊕ ⊕

Lucky's was a real popular joint, a Greek-style place built of brick and shaped like a triangle, with dark windows and the word

Lucky's painted on in red and gold, just like the cigarette. Inside, at one of the high-backed red leather and chrome booths, you could order a greasy cheeseburger, some crinkle-cut fries, and a fountain Coke. Or you could go the whole nine yards and get one of the "specials of the day" that Pete Altieri wrote on the chalkboard hanging over the counter.

Lucky's could've been just another greasy spoon, but I knew it was more than that. From what I could tell, Lucky's was a watering hole for all those men I saw coming in and out of A&W. Every day at lunchtime, there were at least half a dozen guys with thick wads of cash hanging around Lucky's. Some of them looked familiar, like maybe I'd seen them in the paper or on TV. They were a real strange bunch, always whispering, like they were up to something.

Whatever it was those guys were up to, one thing I knew for sure—Pete Altieri was right in the middle. And another thing I knew was that it wasn't his size that got him there. Pete Altieri was a real small guy. Barely five-four. Maybe a hundred forty pounds soaking wet. Throwing his weight around wasn't going to get him anywhere. But he didn't need to do that; for whatever reason, Pete Altieri was at the top of the food chain in this neighborhood.

In some ways I was disappointed by Pete and his friends. They were nothing like most of the gangsters I'd seen on TV shows. These guys were real low key, practically regular. Pete wore a pair of old dungarees and a shirt with rolled-up sleeves and an apron tied around his waist when he was working at the diner. And his buddies didn't dress much better. They all wore sport shirts and slacks. Nothing fancy. No sharp suits. No gold jewelry and pinky rings. I had to figure that if it wasn't a guy's size or his expensive clothes and jewelry that made him the "man to see" in Chicago, then it was something far more complicated than that.

⊕ ⊕ ⊕

As I said, Lucky's was a big teen hangout. It wasn't just mob guys who hung out at the soda fountain. All the gangs went there. One of our rivals had the bad habit of dropping into the joint every so often. Their leader was a big burly kid named Frankie. Frankie was known as a fighter who was always looking to kick some ass. It just so happened that on that particular day, so was I.

Lucky's was packed that afternoon. The place was loaded with girls in ponytails and poodle skirts and guys in leather jackets and ducktails. Buddy Holly's "That'll Be the Day" was playing on the jukebox. Kids were dancing. That joint looked just like a scene out of *Happy Days*.

I wasn't with my usual gang that day. It was just me and two buddies, sitting at a booth in the back, minding our own business. I'd tried to make a little time with a couple of girls and struck out, so now I was getting bored. I started showing off my latest toy—a switchblade. Of course, it wasn't a real switchblade. It was an Italian cheese knife I'd rigged up to scare the crap out of anybody who got in my way. And it worked, too. Like a charm. I had guys running like rabbits the minute I pulled the damned thing. That blade looked real serious. But the truth was, it had never even made its first cut.

About that time, Frankie and his gang came walking in. They swaggered all over the joint, trying to look tough, like they were real bad asses. Pretty soon Frankie and I were calling each other names, going back and forth. When Frankie walked over to my booth with his fist clenched, I knew where things were headed. Evidently so did Pete Altieri, because he came hurrying over and told us to take it outside—which was fine with me.

We marched out to the parking lot, and in no time at all I was kicking Frankie's ass—and kicking it bad, too. It got so one-sided that one of his friends jumped in and tried to help. Now I was fighting two guys. I beat the living crap out of them. Finally they took off running down the street, and there I was, standing by myself in the middle of the parking lot with nobody left to fight. But I still wanted to break some more heads.

That's when I realized that Frankie and his buddies had made a huge mistake: they'd left Frankie's little '55 Chevy. I jumped up on that car hood and went right at it. I kicked out the windshield first. Then I went right on down the line, from fender to fender, from the hood to the trunk. *Bam, bam, bam,* up and down the whole frigging car. I was a one-man demolition derby. I even gave the interior a few major adjustments and used my knife to carve up the leather. When I was done, it was totaled.

By the time I headed back into Lucky's, I felt much better. As it turned out, Pete and his pals had been watching the whole thing, blow by blow. Pete had even taken bets on the deal. If I'd wanted to get their attention, I'd definitely done it.

A few weeks later, I went into Lucky's. It was early and the place was quiet. When Pete saw me sit down in back, he walked over and put a couple of Cokes on the table.

"It's Mike, right?" he said and sat down.

I lit a cigarette and nodded. "Yeah, Mike. Mike Corbitt."

"So you like to rumble, Mike? Get beat up a little?"

I shrugged. "Yeah, I guess so."

He looked me up and down, like he was trying to figure out what made this crazy kid tick, and then said, "You know, Mike, I just don't get it. What's got you so mad that you wanna fight all the time? Hey, I look at you . . . jeez, you gotta lot goin' for you. You know that, kid?"

I nodded coolly.

And suddenly Pete looked concerned. "You got no folks . . . is that it? No Mom and Dad to look out for you? You got no fuckin' money, or what?"

Deadpan, I dragged on my cigarette. "Guess I like to fight. That's all. Guess it just feels good."

"Feels good, huh?" Pete shook his head. "Yeah, sometimes I'd like to smack a few assholes around myself. But . . ." He gave me a big grin. "But I think I better not fuck up this nice face . . . if you know what I mean." He started laughing. Pete had the kind of laugh that made you want to laugh, too. Real contagious.

I looked across the table at him. There was this stub of a cigarette hanging from his mouth. I'd watched him carry on entire conversations with a cigarette hanging from his mouth like that. Even if he was old enough to be my father, I liked Pete. I couldn't help it. I didn't care if he was a gangland criminal or a 1950s entrepreneur.

Right then and there, I dropped the crap. "Yeah," I said. I patted my face and grinned. "Yeah . . . I know what you mean."

"Good," he said, his voice turning serious. "So now we understand each other." He smashed out his cigarette and leaned forward. "I've been lookin' for somebody . . . somebody like you, matter of fact. You know, a kid that can take care of himself, take care of what needs to be taken care of, a kid that can follow orders."

I nodded. I knew what he meant—or thought I did.

"It's for my place across the street, A&W. You know it?"

"Yeah, A&W," I said, trying hard to be cool. "It's an appliance repair company, right?"

"That's right. We're a little shorthanded right now. If you want a job, show up tomorrow after school and you got one. Just tell 'em Pete sent you. Okay?"

"Okay," I said. "I'll be there."

A job at A&W? I thought I was the luckiest son of a bitch in the world. I was going to be there all right. To a kid like me—who didn't know shit from shinola at the time—working for Pete Altieri was like working for the mob.

CHAPTER 6

From the first day I set foot in A&W, I had it knocked. I made seventy-five bucks a week, cash—plus what Pete said I could skim. After a few weeks I had a wad of dough almost as big as some of the guys who hung out over at Lucky's. I was holding two, three hundred bucks at a time. I was on cloud nine. I wasn't even fifteen and I was making more money than my Dad, and he worked like a dog. And what did I have to do for all that dough? It was a real Mickey Mouse job—not at all what I'd thought working for the mob would be like. I didn't have to fight anybody or anything. I didn't even have to steal. All I had to do was ride around in this flatbed truck all day with this cool black guy named Jackie who knew everything about everything.

If Pete Altieri and his friends impressed me, this guy Jackie knocked me out. Maybe it was partly because he was black—at the time, thanks to guys like James Brown, Sammy Davis Jr., and Chuck Berry, being black was the ultimate cool. And Jackie was the king of cool.

Driving around in that junker truck all day, we got to be real tight. Jackie had his eye on the future; his deal was the mob. He was always saying that someday he was going to be one of the guys, a big shot among the Negroes and respected by white people, too. I thought that was a crazy pipe dream, Jackie being black and all, but knowing what I know now, I think he would have made a hell of a crew member.

Jackie was a fabulous salesman; he unloaded more hijacked cigars and swag than anybody I've ever seen. And he had a great memory, too—a bookie's memory. He knew every tavern, strip joint, Italian-American Club, and card game room in the dozen or so communities that made up A&W's South Side territory. Everybody loved him. We got free sodas and sandwiches everyplace we went. At the time, he was my idol, so naturally I watched his every move. He introduced me to the customers and showed me the ropes, but there wasn't much he could teach me about our job— that part was a no-brainer. It was all very routine. Each day we'd get a list of pickups and deliveries and head out on the road. We went from one joint to the next, hauling machines. By the time we got back to A&W it was almost dark. We'd unload the machines and head across the street to Lucky's for a burger and fries.

I couldn't believe they called what we did at A&W "work." And it wasn't, either—at least not until we had to clean the machines. I hated that part of my job. The pinball machines were the worst. Guys would get mad when they lost a game and they'd pour beer, Coke—they'd piss or spit or come on the goddamned things if they could get away with it—and then they'd shove whatever else they could think of down the money slot. Since all the pinball machines were rigged, guys lost at pinball a lot. Which meant that in no time at all the machines were all sticky and gummy and wouldn't work.

When we cleaned a machine, we put the sticky coins in big fifty-five-gallon drums, and Pete would come around and tell us to take a handful for ourselves. I guess you could say I learned the meaning of "dirty money" at A&W. After washing the coins, we hauled them in wheelbarrows to the counters—a bunch of old guys sitting on folding chairs in the back of the shop who sorted the coins with old-style hand-cranked machines. While the counters rolled up the money, Pete's technicians worked on the machines, making sure they were rigged with a sixty-forty payout, which was the way they came to A&W from a mob-owned factory in Kentucky.

As you can imagine, there was money everywhere at A&W. There were always five, sometimes ten, fifty-five-gallon drums lined up along the walls, overflowing with coins. There was change rolled up everywhere, wall to wall. It was like Fort Knox. Of course, I never knew *exactly* what the whole ball of wax was worth at A&W. I knew better than to ask Pete any questions. But eventually I got around to asking Jackie about it.

"So you wanna know what we bring in?" Jackie said. "Hell, boy . . . use your fuckin' head. Add it up. There's four hundred machines out there that have them small boxes. They only hold, what? Seventy-five, a hundred bucks in change at a time, right? And we have to collect on those every other day."

"Three times a week," I said. I had my mental calculator going. "Yeah. So that's a hundred and twenty thousand dollars a week, right there."

"And what about the jar games?" I asked. "We must make a killing on them." Jar games were big glass jars that Pete filled with tickets or pull tabs. Tavern customers paid a few bucks to draw one for a chance to win a hundred dollars. Of course, nobody ever won the big money. Usually they only got a dollar or two. Sometimes we'd hear about a guy winning a twenty, and when word of that got out, business at that particular spot would boom. The jar games were fixed like everything else. Jackie said we probably took in five or six grand a week on jars.

"And then there's the slots," he said. "Slots got their different denominations. Some of 'em hold four hundred a week, some seven or eight. And we got at least four hundred slots right now. That's a hundred sixty thousand dollars . . . on the low side."

Based on Jackie's information, I figured that A&W's total take was somewhere in the neighborhood of three hundred thousand a week, not counting a few miscellaneous expenses, like all those guys Pete had to pay off.

"The cops and officials gotta get their piece of the pie," Jackie said. "Some of the taverns get as much as a third of the take. But most times they get ten or fifteen percent. Plus, there's hundreds of people to take care of, from some motherfucker on patrol to a city commissioner and on up. They all got their hands out. But don't you fuckin' worry. Pete makes out real good here at A&W."

And so did I. I was loaded. Or thought I was. And as it turned out, my job did more than fill my pockets; it also filled in a lot of the blanks I had about the organization I now called "the Outfit." I was seeing men like Frankie LaPorte and Ralph Pierce, who I'd read about in the paper. At the time, Pierce and LaPorte controlled most of the vice rackets on the South Side, including gambling and prostitution, and were bringing in around two hundred grand a day—or about seventy million dollars a year.

Not everyone who walked in the door at A&W was a mob overlord. Mostly it was low-level workers making their weekly pickup. But even if they weren't big shots, I was always respectful. You don't go busting a guy's chops just because he's a nobody you can push around. For all you know he could be the boss someday. Then you'd be screwed. Take me at A&W, for instance. Two of the route guys—Joe Ferriola and Turk Torello—would make it big down the road. Ferriola would be the boss, and Turk, well, he'd get a rep as a stone-cold killer who knew how to get a job done. If I'd screwed around with those guys, my ass would have been in a real wringer when I bumped into them later.

Fortunately, I hit it off right away with Turk and Joe. They didn't play the big shot just because they were with the mob and I was a punk kid. At the time, I figured them for pretty easygoing. Like all the Outfit guys I ever knew, they were nice guys—until somebody screwed them. Then things got nasty. Then those guys could plant you without blinking an eye.

⊕ ⊕ ⊕

I knew something was going on before I even saw the papers. All of a sudden there were a bunch of new stops on our list. Jackie and I were picking up dozens of damaged machines, some burned with acid. Others were all smashed up by a guy called "Sammy the Ax."

Jackie told me there was a push by Sam Giancana and his men to take control of all the machines in the Midwest. To accomplish that,

Jackie said that Giancana had enlisted Hy Larner, a Jewish guy at the top of the Outfit's machine rackets, to put the muscle on the independents. Larner then sent out half a dozen syndicate terrorists to convince honest machine owners to join the mob's newly formed "Amusement Association." It was a pretty straightforward proposition: operators paid a dollar per machine per month to the Amusement Association and its secretary, Hy Larner, or else. Jackie said nobody stood up to Sammy the Ax and lived to tell about it, and after hearing a few of the more lurid stories about the guy, I didn't need convincing.

But evidently there were a few other people who did; over three hundred machine operators in the Chicago area refused to cough up the cash to pay Larner's Amusement Association. Those owners had to learn their lesson the hard way: Larner's thugs made late-night raids, destroying their machines with acid. The newspapers called it the "Acid Wars," and before they were over, the Chicago Outfit controlled all the machines in the Midwest. It didn't seem like much on the surface, but I knew from personal experience that those machines were worth millions.

With a win like that under his belt, I assumed Hy Larner was a major force. But I couldn't be sure of that, because as it turned out, Hy Larner was one of the few subjects even Jackie wouldn't talk about. I heard rumors at work that Larner was "right up there with Mr. Giancana," but that was pretty much it. All the other Outfit guys were in the papers every day, their pictures plastered all over the front page of the *Tribune*. But when Hy Larner's name *was* mentioned in the papers, he was described only as an "associate" or "protégé" of some gangster and nothing more than that. Nobody knew how deep his contacts went or how high up. Reporters called him a "riddle" and a "mystery man." In 1959, when he went before the McClellan Committee, Larner took the Fifth Amendment fifty times. That was the last time the public actually saw the guy. After that there weren't any more pictures. And there weren't any more appearances before a bunch of senators, either. The man the papers called "the Ivy League mobster" disappeared.

⊕ ⊕ ⊕

"One of Larner's guys has a big job for me," Jackie whispered excitedly. Of course, in reality, Jackie didn't have the slightest idea what he was getting into when one of the guys came by and asked him to work late that night. But even if he had, he would have given it a shot; being in the mob was his dream. Later that night we stood out in A&W's parking lot having a smoke and a beer before he had to go pick up his new "partners" at a local bar.

I still remember Jackie's smile as he climbed behind the wheel of his rusty Chevy sedan. "I'm on my way, Mikey," he said as he drove away. That was the last time I saw Jackie alive.

⊕ ⊕ ⊕

I heard later that Jackie was shot dead by a tavern owner. Supposedly he'd been the lookout in a holdup—which I thought was a total fabrication. There wasn't any holdup. More likely it was some idiot who thought he could buck the system and refused to join the Amusement Association, so a couple of Hy Larner's boys had to pay him a visit to convince him otherwise.

Occasionally I'd hear about some machine owner trying to play tough guy. That's when things would go sour. And that's my guess on this deal. When the tavern owner was facing Sammy's ax—or his partner's baseball bat—he must have panicked and pulled a gun out from under the counter and started shooting up the place. Guys like Sammy aren't stupid; when that gun showed up, he and his partner probably ran out the back door and into the alley where Jackie was waiting for them in the Chevy.

Evidently it wasn't enough to run Sammy and his buddy off. That tavern owner wanted to set an example, so he ran after them. He must have been right behind them when they jumped in the car. And there was Jackie, staring back at him through that Chevy windshield. Jackie was probably armed that night. He had a .22 that he liked to carry around; packing heat made him feel important—although I don't think he'd ever even fired the son of a bitch.

He didn't that night, either. No, Jackie didn't pull the trigger and take care of that tavern owner who was waving a gun around, shooting in every direction. Instead Jackie put the car in reverse and hit the gas, and about that time a bullet came through the windshield and hit him right between the eyes.

I imagine it killed him instantly. But Larner's boys probably didn't give a rat's ass whether Jackie was dead or alive. They were too busy trying to get out of there to waste time playing nursemaid to a "colored boy" with a fucking hole in his head. They shoved Jackie right out the car door and took off.

So that's what happened to Jackie. At least that's what I think happened. Part of it I got off the street. The rest I put together on my own. And I thought a lot about it, too. Yeah, I thought a lot about Jackie.

CHAPTER 7

Things were different after Jackie's death. Nobody talked about him at A&W. It was like he was never there. I kept right on working, like nothing happened, but the place didn't feel right anymore. With Jackie gone, it was just me and the technicians and a bunch of old farts cranking some change counters. I was starting to think about quitting. I'm sure Pete noticed I was down. In any case, he cornered me in Lucky's one night. He never mentioned Jackie. He just talked about what was important in life. How if a guy was going to make it in this world he had to get his priorities in line. He said you couldn't afford to let certain things get in the way, that you had to keep your eye on the finish line.

My own father could never have straightened me out the way Pete did that night. I suddenly realized that if I wanted to get ahead in Pete's world, I was going to have to make some changes. I promised myself I'd never let my emotions get in the way again, not when it came to business.

You'd think that after that talk with Pete I would have been right on board at A&W. If life was about making money, I was making it. The only problem was, I didn't think I was making enough. Just seeing all that money come rolling in every day at

A&W made me want more. But I didn't quit my job. I just went and got another one, working nights after I left A&W, setting pins at a bowling alley right down the street.

With two jobs, it was pretty hard to devote any time to school, but I didn't give a rip about getting an education. I was fifteen, and I'd been making more money than my dad for over a year. The way I saw it, an education couldn't buy you a cup of coffee. And when it came to college—well, I couldn't see why anybody would want to go through all those years of school.

I figured I would make it on the street, not behind a desk. Since my gang days, it didn't matter whether I was on foot or behind the wheel, I'd always felt good on the street. Although I didn't have my license, I'd been driving ever since I hooked up with Roger Douglas and his friends. Sometimes it was a car Roger hot-wired, other times it was some kid's parents' car. Every chance I got I was behind the wheel.

The first time I dragged, it was like sex; I got out of that car and I wanted to do it again. I got into street racing big-time. I got off on the challenge and competition almost as much as on the speed. My timing and reactions were first-rate and my instincts were always on target. I was racing just about every night. And I was getting good. In fact I was winning tons of money; three thousand bucks was the usual purse.

My folks didn't have the slightest idea I was out burning rubber up and down the area highways. All the guys that liked to drag went out to DuPage County where we had a mile and a half stretch that was as straight as an arrow. We set things up real professional. We marked the road. We painted lines on it. It was a great time.

If I couldn't find anybody to race, I took on the police. I messed with a couple of cops I should have left alone. But at the time, the name Bill Hanhardt didn't impress me. Hanhardt was one of America's original shoot-first-ask-questions-later cops. To us kids, he was nothing but a swinging dick—the type who likes to throw his weight around. We had no respect for that type of guy. Cops like Bill Hanhardt and his buddies hold a grudge. And to make things worse, they've got nothing but time. If they want you bad enough, they'll get you—which was something I had to learn the hard way, on one particular night when I was out cruising with my buddies.

We'd been driving around with a case of beer in the car, looking for action, when we saw a few girls hanging around a pool hall and pulled over. I was making a little headway with one of them when a squad car came up behind us and two cops got out. I recognized them right away; they were a couple of Bill Hanhardt wanna-bes who'd been looking for me. I'd given them a run for their money on several occasions.

I'd always managed to get away, but not this time. With beer in the car, me and my pals were screwed. When we refused to tell the cops where we got the beer, one of them held a gun on us while his partner took us, one by one, into an alley and roughed us up. After that, I was on the lookout for those bastards. I wanted to get them back, and as luck would have it, not long afterward I came across one of them sitting alone in his squad car in an alley—which was perfect. That cop must have shit his pants when I wrestled his nightstick away and started choking him with it. I dragged him into the backseat and beat the hell out of him. After that, the cops left me alone. They knew I was trouble.

My reputation as a troublemaker wasn't just growing with the cops; it got so bad that regular people got up and left when I walked into a joint. I didn't fight at Lucky's anymore, out of respect for Pete, but that didn't stop me from taking on some smart-ass someplace else. Juvenile officers were constantly dropping by the house. There was always some do-gooder coming over, trying to talk some sense into the "wayward Corbitt boy." None of them had any impact on my life.

But then an Irish priest started dropping by. Somehow he convinced my parents that I would make a terrific man of the cloth. Nobody ever asked my opinion on the matter—if they had, I would have told them straight out that in my entire life I'd never considered being a priest. And what about that vow of poverty? Were they kidding?

I knew it wasn't just having an ear to the Big Guy that appealed to my folks. What really got their attention was that they'd be sending me twenty miles away to the illustrious academy of Saint Procopius, in Lisle.

I screamed and kicked the whole way there. I even shed real tears. But it all fell on deaf ears. The dropped me off at Saint Procopius and that was that. They'd had enough.

⊕ ⊕ ⊕

There aren't any words to describe how much I hated Saint Procopius. There were five hundred students, and just about every one of them was a Bible thumper. I lay awake nights dreaming up ways to get out of there. I did everything I could think of to get under those Jesuits' skin. But all I got were beatings, a bunch of detentions—and a rep as a hard-ass. By late spring, I'd made up my mind that I was running. I became a model student, which meant I was off detention and got to go home on weekends. For a while I played it straight. I went home on Friday afternoons and returned to Saint Procopius on Monday morning just like I was supposed to. And then one weekend when I thought the time was right, I left and didn't go back.

My folks thought I'd gone back to school that Monday morning. The priests just figured I was sick. I knew I'd get caught eventually, but until then I intended to enjoy myself. I ran into my old pal Roger Douglas at a bowling alley and we started hanging out in his '51 Chevy. At least I thought it was his. I had no idea the car was stolen.

After a few days of cruising all over town, we started getting bored, and Roger suggested we could rob a few gas stations. I hadn't done a break-in before, so the idea really turned me on. When I had second thoughts, I reminded myself that we weren't doing *armed* robbery and were going in after closing so there wouldn't be anyone to worry about. It would be easy.

And that's exactly how it was, too. We knocked down three, maybe four gas stations without a problem and then decided to hit a Standard station on Blue Island. We got there around two o'clock in the morning. It was real quiet. I pulled into the alley and waited in the car while Roger went around front. Pretty soon I heard glass breaking. Roger was making so much noise that I started worrying. Maybe we'd screwed up by picking this particular station; maybe things there were *too* quiet.

That's when I saw a car coming down the street toward us, no lights on, moving real slow. Right then I knew there was a silent alarm in the place. We were screwed, and all I could do was sit there and watch. Pretty soon that gas station lot was full of cops. Then a car drove up and parked at the end of the alley, blocking me in. Now I was a sitting duck. I couldn't wait for Roger; I bailed out and took off running. I only got about a block before a couple of cops grabbed me. They threw me in the backseat of a squad car and we headed back to the gas station.

As it turned out, Roger had seen those unmarked cars, too. But he didn't take off on foot; instead he jumped in the car and rammed through that cop car blocking the alley. He sped out of there like a bat out of hell, and the car I was in was right behind him, with me handcuffed to an eyebolt in the backseat.

It turned out to be a hell of a high-speed chase. Roger was barreling through barricades, crashing through roadblocks, there were cops scrambling in all directions. We were doing eighty, maybe ninety miles an hour, and I was knocking around the backseat like a pinball. Every turn was a hairpin.

About the time things started getting scary, we skidded around a real sharp corner, and there was Roger's Chevy. It was upside down, smoking, its wheels spinning. Like a flash, Roger was out of the car. And the cops were right behind him, chasing him on foot. The next thing I heard was a loud crack, then another. Roger was unarmed, but that didn't matter; they were shooting at him.

Fortunately, he got away. It gave me a lot of satisfaction knowing they didn't get him, but as I found out later, it ate at those cops, thinking that some kid had given them the slip like that. They might not know his name, but they knew what he looked like. And from then on, they'd be gunning for Roger Douglas.

After they gave up on apprehending Roger, the cops hauled me over to Blue Island and threw me in a pen with a bunch of hardcase killers and thieves. I couldn't convince them that I was only sixteen and didn't belong in the same cell as murderers. And they didn't believe that I didn't know my friend's name, either. For that, I got a good solid beating. By the time they were through with me, I was begging to call my folks. I was prepared to take whatever my dad could dish out just to get out of there. But they wouldn't let me call my parents; instead they took me down to the county juvenile hall. By that time I was scared. But when I walked into the courtroom and saw Dad sitting next to a Summit policeman, my fear turned to shock.

Before I knew what had happened, the judge had turned me over to the Summit cop, and for a split second I thought I was getting off scot-free. But when they slapped cuffs on me and took me out to the squad car and shoved me inside, I knew it was just the beginning. Dad climbed in, too, but he wouldn't look at me. He wouldn't say a word. It was a long ride to Summit.

We went straight to the police station and they put me in the lockup. Nobody would tell me anything. I stayed there for a couple of hours, wondering what was going on, until finally they took me to the chief's office, where Dad and a judge were waiting. The judge told me I could go to trial and get a sentence or I could go into the army. Those were my choices. The military or jail. There was no way in hell I was going to jail so it was settled; I'd join the army.

A few days later I went to enlist, and they turned me down flat—a guy six-four, two hundred pounds, I couldn't believe it. They said I'd failed the hearing test, that I had a birth defect. Since the army wouldn't take me, I tried the navy. But they wouldn't take me, either. Finally I went to the marines. By this time I knew how to beat the hearing test. The guy said, "Do you hear this? Do you hear that?" I just said "Yeah, yeah, yeah" to all his questions and passed the test. Now I was a marine.

I went home and told my folks that I was leaving right away. My mom cried the whole time she packed my bag. Pretty soon the entire family was crying along with her. They took me down to a city bus and we said good-bye. It was real sad. We all cried like babies.

The bus took me to debarkation, where everybody got reexamined. This time when I took the hearing test I flunked. They told me to see a specialist. I got back on the bus and headed home. My folks were real unhappy when they saw me standing at the door. The very next day, Dad marched me right down to Taylor Street to a hearing specialist. The specialist confirmed what I'd been told. So that was that. No military. And no jail time, either. I was off the hook.

I think the fact that I got rejected bothered Dad, but not me. I figured that I was supposed to stay in Chicago, that I was cut out for better things in life than marching around in some uniform. What the hell, maybe I was even headed for greatness.

CHAPTER 8

f I was marked for great things, it wasn't evident by the time 1961 rolled around. I admit that a lot of it was my own damned fault. For one thing, I never finished high school. After Saint Procopius and my little military fiasco, I went back to public school, but I got kicked out and that was the end of it. Not a real bright move—but I was one of those smart-ass kids who think they know it all.

In my own defense, after I left high school I didn't just lay around my folk's house waiting for a free meal. I got a job to help out with the expenses—and to keep me in the street-racing game. But it didn't take a genius to figure out it wasn't going to work with me living at home. Being somebody's kid wasn't for me—so what if I was only sixteen. I was too independent. I had to move out.

I got a place of my own. I already had two jobs. Now I got three. Next thing I knew, I was working around the clock. I even worked at a factory making car seats—which was about the time I started getting worried that I was turning into my father, that my destiny wasn't going to be anything more than punching a time clock.

So there I was—it was 1961 and I was seventeen years old and I was out on my own, with nobody to blame but myself if things didn't go right. It was a real strange time. Even my love life was taking a weird turn. Yeah, Mike Corbitt—king of the first date score—had gone and fallen in love.

⊕ ⊕ ⊕

I'd dated tons of girls before Annette Vogel. I practically had to beat them off me. Back then, a guy who could dance had it made. Most girls were easy when you had the right moves; shoe leather really turned them on. But this girl, Annette, was a whole different deal. She was one class act. She wasn't a big dancer. And she wasn't easy.

When I think back, I realize that, as beautiful as she was, Annette Vogel was just a shy, lonely little girl. Maybe that was what I dug about her. She didn't come on to me like all the others. She was reserved and proper. And as square as you could get. The first time I saw her I'd just come off my Saint Procopius escapade. She was standing on the sidelines at a dance, looking real cute, very Jackie Kennedy, with this big blond bouffant hairdo. She had on a pink mohair cardigan buttoned up real tight. She was so stacked I couldn't take my eyes off her.

I decided right then and there that Annette Vogel was going to be mine. There was just one problem: she was with another guy, a real jerk. But I wasn't about to be put off by competition, especially not by him. I figured that before the night was over the guy would screw up somehow. Sure enough, that's exactly what happened. After the dance, he started getting pushy with Annette in the parking lot. And there was my opening. I just stepped in. I could tell the guy knew my rep because he took one look at me and hit the road. I got a real kick out of that. Evidently so did Annette, because that night I took her home.

⊕ ⊕ ⊕

I never felt like anybody cared about me before Annette Vogel. Deep down I figured I wasn't worth much more than a good fight. But as far as Annette was concerned, I could do no wrong. She idolized me. I was older, and I was the coolest guy she'd ever known. Plus, I had money and my own place and a car. But more important than that, I was a tough guy—which meant I could protect her. And she needed protecting, too, especially from her old man. Annette's father was real rough on the whole family. Naturally, he hated me from the start. But that didn't stop me; I was crazy about his daughter.

I was so crazy about Annette that, even though I hadn't seen the inside of a school in ages, I rented a tux and went to her senior prom. I planned this real romantic night with dinner at the Villa Venice after the dance. At that time the Villa Venice was all the rage. It was a mob-run joint and was very plush and *very* Italian. The place even had gondola rides. And there was always big-name entertainment on stage—Sinatra, Martin and Lewis, Louie Prima. Some nights you'd see Sam Giancana and his rat pack friends sitting at a table up front by the stage, all of them looking real ritzy with a bunch of gorgeous broads hanging on their arms. Obviously, the Villa wasn't a joint where high school kids hung out. It was very sophisticated and *very* expensive. It almost took me broke taking Annette there. But I was in love, and because of that I would have spent my last dime.

It was a terrific night all around, but what put it over the top was the gondola ride under the stars after dinner. Talk about romantic. Annette was kissing me and whispering in my ear that she loved me more than anything in the whole world. And I felt the same way about her. I was in love all right. But that's as far as it went for me. When it came to a lifetime commitment, forget about it. And as it turned out, that's exactly what Annette wanted. There we were, riding along in that gondola, and all of a sudden she looked up in my eyes and said, "Let's promise each other that we'll always be together . . . forever."

Forever? Normally that word sent me running. But that night I just looked into Annette's pretty eyes. Maybe it was the way she said it. Maybe it was the romance of being under all those stars. Or maybe it was that six-pack we put down on the way over. Whatever it was, I didn't disagree. I nodded and said, "Forever," and kissed her real hard, like that would make it all come true.

I have no idea what I was thinking. But that night was the beginning of the end for us—even though the end was seven years away. All that *forever* crap took us places we weren't ready for. I see now that it wasn't just me that screwed up on the deal, either. Annette probably wasn't in love with me. She didn't want commitment. She just wanted out of her father's house. And who could blame her? In fairness, we were both too young to know the first thing about love.

In any case, logic didn't enter the picture; a few days after that gondola ride, the word *forever* turned into *eternity* and Annette started pressuring me to marry her. She said that if I didn't, she'd go away to school and we'd be over. She started demanding to know what I wanted out of life, and that's when I realized that I needed Annette Vogel just as much as she needed me. I also realized that I didn't know what I wanted out of anything—let alone my life.

The truth was, I was real down about how things were turning out. When I'd moved out of my folks' house, I thought that was it— that Mike Corbitt was on his way to bigger and better things. But it hadn't gone like I'd planned. I was working day and night. And all those fancy dreams I'd had? They were still dreams. But the worst part was that I was starting to think I was becoming my father. And that scared the hell out of me.

Over the next few weeks Annette kept the pressure on. And I kept thinking about my father. I started wondering if all those dreams I'd had would ever come true if I didn't have her by my side, cheering me on. Talk about crazy.

We had an old-style Corbitt wedding with the whole shebang in church. Annette became a Catholic, which made my mother happier than I'd ever seen her in my whole life. Dad was pleased as punch about the deal, too, mainly because he figured that now I'd finally settle down and his worries would be over.

After the reception, Annette and I headed south to the family farm for our honeymoon. Twenty-four hours later I took a hard look in the mirror and—*bam*—talk about a major wake-up call. I knew I'd really screwed up this time. I was still a kid. I didn't have the slightest idea what being a husband was about. And what did Annette know about being a wife? We were going to play this one by ear—which isn't a good way to start anything that's supposed to last *forever.*

We hardly got back home and settled into our new garden apartment in Summit when I let everybody—including Annette—know just how I felt about all that matrimony crap. I wasn't abusive or anything. I didn't cheat on her. I just stayed away. I was never at home. I worked like a Trojan, all the time, day and night. And when I wasn't working I was out on the streets, racing cars.

Before I knew it, two years had gone by and it was 1963. I was still married to somebody I didn't love. And I was still working my ass off. Only this time it wasn't at some car seat factory. This time it was at a gas station. I was moving up the ladder all right. Where was a break when you needed one?

⊕ ⊕ ⊕

Pete Altieri and my boss, Tony Pappas, went way back. They knew all the same people, had all the same friends. Which was how I'd landed the job at the gas station to begin with. Like they say, it's who you know that counts. Pete Altieri vouched for me and that was that, I was working for Tony Pappas.

Getting the job at Tony's station wasn't exactly a big move up the ladder. I was nineteen years old and a gas station attendant. But it *was* money. And with a wife and a bunch of bills to pay, I figured I'd better get with the program. Then again, it wasn't like Tony's place was your typical grease monkey gas station. Forget about lube jobs and oil changes—there was all kinds of kinky shit going down in that joint. It didn't take me long to get the lay of the land, which in this case consisted of Tony's day-to-day business with his Outfit buddies. Some of them I'd seen before at A&W—Turk Torello and Joe Ferriola, to name two. Others, like Rocky Infelice and Tony Spilotro, I got to know over time.

There were also a lot of mob wanna-bes coming around the station. They were young and spent most of their time trying to impress the big boys. Some of them—like Billy McCarthy and Frank Miralgia and my old friend from the neighborhood Roger Douglas—lived right on the edge. Forget about routes and machines and bookmaking. These guys packed heat and did holdups.

I had to admit, it paid well. If a guy was smart and hit the right joints and paid off the right people, things went along smooth as silk. The Outfit guys and the cops left you alone, and you could make ten, maybe fifteen grand a month. That's if a guy was smart. Evidently Miralgia and McCarthy weren't the world's greatest geniuses, because in 1962 they went and nailed some people in a suburb of Chicago called Elmwood Park—an area that was strictly off-limits—and on top of it they did it without getting permission. That's the sort of crap that gets a guy killed. It sure as hell ended their careers.

Everybody called it the M&M murders—because of Miralgia and McCarthy being their names—and for months afterward, all you heard was how Tony Spilotro put Billy McCarthy's head in a vice and turned it so tight the poor son of a bitch's eyes popped out. Not a pretty picture, but that's how Spilotro made his bones.

Spilotro had a small-guy complex. Power is what tripped his trigger. He got off on making some poor bastard beg for his life. And it was that attitude that moved him up the ladder; after the M&M murders, Spilotro's career really took off.

The M&M murders had an effect on my career, too—they made me start thinking about where I wanted to go in the organization. And how I was going to get there. It always came down to the green. I wanted to be on the inside, but I knew I wasn't cut out to be an enforcer. To be an enforcer a guy had to have no heart at all, forget about a conscience. And that included guys I knew and liked, too—guys like Turk and Joe. They might seem like they were totally different from Tony Spilotro, but they were all cut from the same cloth. I'd heard the horror stories—the ones about them and their boss, a guy the papers called the Outfit's "master of juice collection," Sam "Mad Dog" DeStefano.

DeStefano wasn't just into the juice. From what I'd heard he had his fingers in a dozen different pies, from bookmaking to counterfeiting. Evidently he also had a special place in his heart for stick-ups, because he bankrolled just about all the major robberies. He was also into hijackings and high-class burglaries—jewelry heists, furs, stuff like that.

With those kind of big-ticket hobbies, you might get the idea that our man DeStefano was just another Outfit guy with a weird nickname and good taste. But they didn't call him "Mad Dog" for nothing. Sam DeStefano was a real nutcase who actually foamed at the mouth when he did somebody. He got his rocks off using an ice pick. He'd even built a soundproof torture chamber right in his basement. Nobody wanted a tour of Sam DeStefano's house.

Since I'd been out on my own, there'd been dozens of real grisly murders in Chicago—and not one had been solved. But the Outfit guys knew who was behind them all: it was DeStefano and two of his favorite enforcers, Chuckie Nicoletti and Phil Alderisio. The word on the street was that the cops knew who the perpetrators were, they just turned their heads. And really, they had to know; those murders had Mad Dog's name written all over them. The victims weren't just shot in the head and stuffed in some trunk. That would've been too kind—and probably too quick for DeStefano's kinky taste. No, the poor bastards were tortured. Cattle prods up the ass. Knives. Meat hooks. Ice picks.

Sometimes Mad Dog and his pals actually took photographs of their victims. One of the worst was the one they took of a hit they did on a loan shark named Action Jackson. Everybody said Jackson was a real sleazebag degenerate, but that wasn't why they whacked him. Supposedly, he'd been talking to the FBI. For that, Jackson was tortured for days before he died.

Every so often somebody would come by Tony's station with pictures of Jackson, all sliced and diced. Talk about sick. The guys would pass those pictures around, and pretty soon they'd start up with the jokes and gory details. One thing about those old-time Outfit guys, they just loved to talk about a hit. It was like they got to live it over and over again.

⊕ ⊕ ⊕

My boss, Tony Pappas, and his brother, Cookie, were nothing at all like Sam DeStefano and his boys. The Pappas brothers were real regular guys, the next-door neighbor type. But then, like I've said before, you don't have to be a killer or run around breaking guys' legs to enjoy the advantages of being connected to Outfit guys. If they make money, you make money. And that's how it was at Tony's station. Just like at A&W, the joint was a money machine. It was one sweet deal after another. Some of their best deals were very straightforward—the old tried-and-true gambling stuff. Like when one of the guys would pay Tony to close the station for the weekend and there'd be some big card game that went around the clock.

Even though Tony and his brother made money hand over fist, they weren't actually what you'd call "players." When it came to mob activities, they were middlemen—which isn't such a bad spot to be in, really. That way you're not so big that you create rivalry—which always leads to trouble—but you're not so small that nobody gives a rat's ass if you end up in the river.

There are a few other advantages to being a middleman, too—like not getting a lot of heat. In all the time I was at that gas station, Tony and his brother never got pinched. If the cops came around, Tony just paid them off. Occasionally Tony and his brother would fence for Turk and Joe or a few of the Melrose Park boys. One morning you'd go in and there'd be a truck or two parked out back, behind the chain-link fence. Usually it was a load of hijacked cigarettes. Sometimes it was clothes or radios or TVs. Whatever it was, by noon the word was out and all the guys were showing up, looking for their piece of the action. You would've thought they'd won the lottery the way they acted. The meanest son of a bitch was all smiles. And it didn't matter if the haul was big, either. What mattered was they were getting something for nothing.

After working at Tony's station for a while and getting to know most of the guys, I figured some good opportunities were going to come my way. But months went by and there was nothing. It was very discouraging seeing all that money change hands and not one damned dime of it come by me. Anybody else might've gotten fed up and quit, but I was determined to make it. All I needed was just one break. I had to believe it was only a matter of time.

⊕ ⊕ ⊕

Usually when Pete Altieri came by Tony's place, we'd shoot the breeze, maybe have a Coke and a cigarette, but on this particular day there was none of that. From the minute Pete got out of his car, he was all business. He didn't beat around the bush. "You ever thought about runnin' your own station?" he said.

I told him, yeah, sure, that I'd thought about being my own boss lots of times—but I was lying. I'd always figured owning a station was way out of my league.

Pete went on, "What if I told you there's a station available? A Sunoco. A real beauty, too. A guy by the name of Carl Smith was running it. Seems he got into a problem at the track, so now the guys want him out. They're lookin' for somebody with a good head on his shoulders to run the joint. Your name came up. If you want it . . . it's yours. All you gotta do is pay seven grand for the first gas delivery and you're on your way. So what do you say? We got a deal?" Pete stuck out his hand and grinned.

Seven thousand dollars? It might as well have been a million— that's how close I was to having that kind of dough. But I had to go for it. I knew that station. It was a beauty all right. It was painted up real pretty in Sunoco colors: yellow and blue and white. It had two fenced back lots and four bays with three utilities. There was a wash and three hoists. Two rows of pumps. It was way bigger than Tony's joint. There was no telling what I could bring down at that Sunoco.

I shook Pete's hand. "It's a deal," I said, like I didn't have a care in the world. But a few minutes later, when he drove off, reality hit. I didn't have the slightest idea how I was going to get my hands on seven grand. All I knew was that I had to—somehow. This was my chance, maybe my only chance. Lady Luck had just dealt me a hell of a hand. I had to play it.

CHAPTER 9

Before I opened the door to the Sunoco, I stood out front looking at the joint. I couldn't stop smiling. It wasn't that long ago I'd been breaking into gas stations. And now I owned one. Wow.

It hadn't been easy coming up with seven grand. I scraped together every dime Annette and I had, which wasn't much, and then went looking for more. If my folks hadn't kicked in the rest, I probably would've done something drastic. I had to have that station.

I was very nervous about taking so much money from Mom and Dad. They'd loaned me over six thousand bucks, all they had in the world. I couldn't stop worrying about how I was going to pay them back. And that wasn't all I was worried about, either. After I got through paying for the first gas delivery and got all my supplies lined up, I was totally broke. I didn't even have my next month's rent.

Fortunately, I had a few things going for me. For one, there was Pete. If I had a problem, I knew I could go to him. For another, thanks to my job at Tony's, I knew all the angles of running a station. But I didn't want to rip people off; I wanted to build my place into a neighborhood station where regular folks knew they'd be treated right. I was one determined son of a bitch, too. I wasn't going to rely on Pete and his "friends" to make the station a success. I started rounding up corporate accounts. I must have knocked on a thousand doors, and in just a couple of weeks I had executives from all the area factories lined up, begging for a tank of gas. Next I went after the local cops. Talk about balls. These were the same guys who'd chased me all over town a few years before, but forget all that; somehow I persuaded them to bring me their business.

For months I worked around the clock, night and day. Halfway through the first year, Pete threw a little surprise party. Nothing much—a bottle of champagne, some beer, and some broads, a few of the guys from the inside, like Turk and Joe. Pete told me the guys upstairs wanted me to know they were real happy with the job I'd done.

And I *was* doing a great job. I'd been getting raves from the Sunoco people. Before I came along, the station's biggest month was ten thousand gallons. Now I was pumping thirty thousand gallons *a week*. After I paid all the bills at the end of the week, I still had a thousand bucks in my pocket, all legit. Of course, I knew a dozen other ways to make money that weren't. I hadn't even started to scratch the surface.

⊕ ⊕ ⊕

Once you've got your feet wet in a particular business, making money on the side is strictly a matter of being creative. In a gas station, cash repair jobs can line your pockets. You do the work, but you don't run it through the books. Or, instead of buying all your parts and oil through the company, in my case Sun Oil, where you're invoiced and everything's real up and up, you buy a few things on the side, with cash, through some cut-rate joint. Then you turn around and sell them—for cash. Nothing goes in the register. Forget about Uncle Sam. After a few months I was pulling in over eight hundred a week on that trick alone.

Between the legitimate business I was doing and the kinky stuff I had going on the side, I was flying. That first Thanksgiving I paid my folks back and never let them buy a gallon of gas again. At Christmas I bought Annette a fox coat. By spring we had an apartment full of classy furniture and our closets were crammed with new clothes. In the beginning Annette had been very skeptical about the gas station business, but by the time we got a Corvette that spring, she was on board—she even pitched in and did the station's books. For the first time in our marriage we were a team. We even started talking about our plans for the future and having a kid.

It might sound like I was settling down, but there was no way that was going to happen. In fact, now that the gas station was such a moneymaker, I was able to give my rebellious nature full rein. I got back into racing big-time, driving professionally. I was winning, too. I had the world by the ass.

Nobody ever paid the tab when I was around. I bought a new Corvette every year after that. I never hurt for money again. And that's what mattered most to me. The almighty greenback. I thought there was nothing it couldn't buy. Ten years later I'd have a house in the burbs worth half a million bucks, and stacks of hundred-dollar bills everywhere I looked. We're talking big money here—stacks eight inches thick, stuck in hidden panels all over the place. In 1987, when I got pinched, I had thousands of dollars in my pocket. And what did all that money mean? Not a thing.

Funny how you never get the message until they slam that steel door on your ass. But enlightenment doesn't come cheap. And as far as freedom's concerned? Well, let me tell you something about freedom—when you're behind bars, freedom is fucking priceless.

$$\oplus \quad \oplus \quad \oplus$$

I never knew if the Outfit guys were instructed to trade at my Sunoco or not, but from the beginning all the old gang from Tony's station was coming around. Besides Pete, who was always there if I needed him, it was mostly the route guys and juice collectors. Turk Torello and Joe Ferriola were at the Sunoco all the time—cutting up, drinking Cokes, talking on the phone. They had a vending route, so naturally they loaded me down with a bunch of machines. But Turk and Joe weren't heavy into the machine racket. Their deal was juice collecting. Vending was just a cover so they could drive around all day and beat up on guys who owed them and their friends money.

Turk and Joe were always on the pay phone at the station, calling some poor slob, threatening to break his head. It was scary, hearing the way they talked to people. These weren't the same guys I knew from my old days at A&W. So when Turk and Joe brought this guy named Mugsy Tortorella to see me and said they were looking for a place to stash a few hot trucks and trailers, I wasn't exactly inclined to turn them away. And then there was Mugsy; his reputation preceded him. Supposedly, Mugsy Tortorella would do anything for a buck. Anything. He had a rap sheet as long as my arm. He was a bookie, a burglar, a home invader, you name it. From what I'd heard, he'd even been involved in Marilyn Monroe's death—which according to the guys was a contract job, not a suicide. So with Mugsy, it wasn't like murder was out of the question—not as long as the price was right. It all made sense once I found out that Mad Dog DeStefano was behind Mugsy's hijacking operation. Killing wasn't murder to DeStefano, it was entertainment. Any guy tight with Mad Dog had to be a stone-cold killer as well.

So from the minute we all went in the backroom and they started giving me the pitch, I had no desire to have Mugsy Tortorella hanging out at my Sunoco. But I had to play along. And the more those guys talked about their hijacking operation, the more I didn't want to go along with the deal. Mugsy's hijacks sounded risky—not at all like what Tony Pappas had been doing. This wasn't just cigarettes and golf shirts.

I tried to stall for a while, but I could tell they wanted an answer. They were getting impatient. Finally Turk looked over at Joe and Mugsy and they got up and left. At that point I figured I had a real problem. Fortunately, Turk was a straight shooter. He said, "Hey, forget that bullshit you heard about Mugsy. Don't you worry about nothin'. He'll take care of the cops. And pay you real good, too. Five hundred a fuckin' load, more if it's a big one. You got my word on that, Mike. And if things go nice and smooth, all the good loads will come by you. You'll be rollin' in it."

Rolling in it? When Turk said that, all my concerns about those guys hanging out at my station went out the window. Suddenly I was thinking about that dough—five hundred bucks a load, maybe more. I told Turk to bring it on. We shook hands and the deal was done.

⊕ ⊕ ⊕

You could always tell when a new load of Mugsy's swag was in town. Like the time the guys got their hands on a truckload of Coppertone's Sudden Tan. Every mafia princess in Chicago was orange for a month. And at Christmas, when those troll dolls were big, every Outfit guy's kid had a bunch of those ugly little suckers under their tree. When there was a shipment of Polaroid cameras, you couldn't go anywhere without running into some asshole taking pictures of everybody. And then there were the pantyhose. When they came out, there wasn't a wife or girlfriend who didn't have a dozen pair.

Like everything else, the hijack business was booming at my station. Sometimes Mugsy's guys brought in two or three loads at a time. I always made a pretty good score on top of my regular cut. You know, "Here's a TV, a mink coat. . . ." Before long Annette and I had all sorts of crap. Everybody came out good on Mugsy's hijacks. And forget about stickups. Nobody's life was ever on the line. Nobody got hurt. Nobody got pinched. There was a lot of cooperation in Chicago, not at all like the Five Families in New York.

Mugsy said those guys were nuts. "Those New York greaseballs got fuckin' bodies flyin' in all directions," he said. "There's no control in their organization. No fuckin' honor, either. They're all assholes, that's their problem. And the sons of bitches are greedy, too. Hey, let the other guy wet his beak a little, what the fuck do you care? You don't get much cooperation doing business like that. But do those guys out east give a rat's ass about that? Fuck no. They'd rather whack some poor cocksucker and keep his cut for themselves. Not me. I got no problem spreadin' the wealth."

Mugsy was a hell of a businessman; he ran a tight ship. Once a load of swag got to my place, things moved like clockwork. The word went out that a load was in, and the fences marched right on over. They'd load up their vans and head off to construction sites, groceries, liquor stores, clothing operations—the list of customers went on and on. Guys actually specialized in specific types of stolen goods. That's how big the industry was. And every deal that parked in my lot, I got a cut. Five hundred here, a thousand there. All cash. Like Turk had promised, I was rolling in it.

⊕ ⊕ ⊕

When there's a new guy on the block, the cops always want to find out if he's an easy mark for a shakedown. When my Sunoco opened, they were just looking for some excuse to put the squeeze on. Not that an excuse was hard to find when a guy had a back lot full of swag. So it was inevitable that the cops would come storming into my joint, looking for a few extra on the side. That was their job. My job was to make sure they got as little as possible.

Over the months I must've been raided a dozen times by the local police, but even though they tried to clip me for a grand or two every time, they never walked out with more than a few hundred. Things were running pretty smooth, and then, out of nowhere, the hammer fell. This time it wasn't the local cops that came down on me. This time it was three carloads of plainclothes hijacking crimes detectives. Busting guys like me wasn't their hobby. It was their business.

When those cops hit my joint, you would've thought they were a SWAT team. Forget about *Dragnet*. Those guys didn't waste their time getting the facts. They made a beeline for the hot truck behind the chain-link fence. I just stood there watching, wondering whatever happened to all those bribes and payoffs they were supposed to get before a load got to my place. Obviously, something had gone wrong on this deal. When one of the detectives walked over, I figured this was it. But instead of cuffing me and hauling the truck off, he asked if they could take a look inside the truck. What was I going to say? I relied on my old tried-and-true technique: I played dumb. I told the detective, "Sure, help yourselves."

At the time, it didn't occur to me they'd take that "help yourselves" remark literally. But they got that truck open and it was like an all-you-can-eat buffet at a hog festival. As it turned out, the truck was full of men's suits. Rows and rows of crates of Hart Shaffner & Marx. Real class swag. Those detectives took one look, got on the horn, and in no time at all there were thirty squad cars parked at my station.

When the detectives opened that truck, it was full. Eight hours and a few hundred suits later, when they towed it away, it was half empty. Those rat bastards cleaned it right out. Slick as a whistle. That was the end of the line for me. Even though I didn't get pinched, I'd had it. I called Turk and told him I didn't want any more swag coming by my station. Maybe it was the way I said it—I was pretty hot at the time—but I didn't get any argument. Turk said if I wanted out, it was my call.

Once I cooled down, I started getting worried. After all, it was those guys who had set me up at the station. They could just tell me to go fuck myself and keep bringing their loads in—what was I going to do? Or they could get somebody else to come in and take over. I'd heard stories about guys trying to get out of Outfit deals. Sometimes there were serious repercussions.

Fortunately, in my case nobody got bent out of shape. It was just over. Although I didn't know it at the time, there was a good reason nobody gave me any trouble. Of course, with those guys there's always a reason. As it turned out, Turk and his friends had bigger things in mind for my Sunoco.

⊕ ⊕ ⊕

It might sound like I did everything *but* run a gas station, like it was all greasing palms instead of axles. But my legitimate gas station business was growing. In fact, it got so busy I had to hire a few mechanics to help out. I was working around the clock. Over time I'd gotten to know all the other race car drivers, and they'd started coming into the station looking to use one of the racks for one thing or another. And then there were the local mom-and-pops that needed a tune-up and an oil change. And of course, Turk and Joe and their buddies were bringing their cars in, too. They were always wanting something done. They all knew I'd take care of them—which was good because when you do those guys a favor, they know how to return it. They let their pals know you're somebody they can do business with. That's when things start to happen.

$\oplus \quad \oplus \quad \oplus$

My first impression of Marshall Caifano was just total amazement. He was a very sharp dresser. He drove a real nice car. And the broads he had around him were absolutely beautiful. You immediately knew he was somebody special. You can't imagine how much respect he got; it was like he was a god. I understand that for a while, that's just about what he was, too—at least in Las Vegas.

In the fifties, when the Outfit moved Johnny Roselli out of Vegas, Sam Giancana sent Marshall Caifano out west to represent the Outfit's casino interests. Caifano had been tight with Giancana for years. Supposedly, Giancana had personally raked in over three hundred thousand a month from the skim in Vegas when Marshall Caifano was working out there. Word had it that the Chicago Outfit's skim had been in the millions under Caifano. Every month.

But not everybody had been happy with Marshall Caifano's performance. From the minute he hit Las Vegas, the low-profile guys, like the old-time Capone guy Tony Accardo, thought the flashy Caifano would bring down too much heat on their action in the casinos. Nobody said that out loud. Who would have had the balls? Sam Giancana was Caifano's chinaman. But they were right. Pretty soon, things actually started going south. Caifano didn't know the meaning of low profile, and thanks to that, he had the honor of being one of only eleven guys—along with Sam Giancana—who made it into the Nevada Gaming Commission's infamous Black Book, the list of mob guys banned from the Vegas casinos.

By the time Marshall Caifano drove into my Sunoco in his great big shiny new Cadillac with a gorgeous blond snuggled up next to him, he was pretty much persona non grata in Las Vegas. He was living down the street from Sam Giancana in Chicago and running a bunch of companies, mostly restaurants and taverns, around Rush Street.

From day one, I treated Marshall Caifano with total respect. He never paid for a gallon of gas at my joint. At first things were pretty formal between us. But after a while they loosened up and he started parking his car behind the chain-link fence out back on Fridays. I'd have my guys wash it and wax it real nice so it would be all detailed out when he showed up on Monday morning. It was no big deal. But in the Outfit, that sort of treatment goes a long way. Pretty soon, "Mr. Caifano" was "Marshall," and he and his buddies were meeting in my back room on a routine basis.

Before I knew it there were a lot of new guys coming by the station thanks to Marshall. I left them alone, I let them conduct business. I never imposed myself on anybody. I knew that eventually Marshall would bring them around and make the introductions.

There was one guy he brought by—his name was Butch Blasi. I knew he was important, but just looking at him, I couldn't see it. For starters, Blasi wasn't a good dresser. And as far as being a playboy like Marshall? Forget about it. Blasi was a slob. But despite that, it was plain the guy carried some weight in the Outfit.

Not long after we met, Blasi drove into the station by himself. It was late at night. There was nobody around, and I got the impression that was just the way he wanted it. He pulled over by the side, where it was dark, and turned off his headlights. He said, "There's this guy I know. He wants to put his car out of sight for the weekend." He was talking very low, like he was worried somebody might be listening. "This guy don't want nobody going around his car. Anybody wants to touch it, forget about it. You understand?" The way Blasi talked, you would have thought his pal was the King of Siam. But whoever he was, Blasi wasn't giving up his name.

The next morning Blasi's friend dropped off the car, but I never got a look at him. It was another week before he came around to pick up his car—and again I missed him. By this time my curiosity was starting to get the best of me. Fortunately, I didn't have long to wait. That same afternoon, a car pulled up to the pumps and a man rolled down the window and waved me over. Right away I knew it was Blasi's friend. Maybe it was because he had this enormous charisma.

At first I thought he was some big-shot executive who'd been fooling around on his wife. He was very well dressed, in a perfectly tailored gray suit and a fedora with the brim turned down over his eyes just so. His eyes were very dark. And hard, like steel. He didn't smile. He just looked at me with those eyes, stuck a roll of hundreds out the window, and said, "Here's a little something for your help, kid. I want you to take care of my car from now on. And do me a favor, will you? Give Butch your pay phone number. I'll call you now and then. . . . We'll send some business your way."

There I was, watching the guy drive away, and I still didn't have the slightest idea who he was. Turk was standing in the door of the station, grinning real big. He'd seen the whole thing. He said that he wanted to shake my hand, that I'd managed to do something some guys never did: I'd impressed the big guy, Sam Giancana.

CHAPTER 10

After I met Sam Giancana, things started to take off. Mr. G., as I called him at that time, was as good as his word. I started taking care of his car, and he began sending business my way. Most weekends, his car was in the bay. On Fridays he'd show up and wait for Butch Blasi. I'd give him a Coke and we'd talk awhile. As time went on, I got to read him pretty well.

Sam Giancana was very moody. He could give you a look that was next to none. A killer look. When he was like that, he'd just sit in his car saying nothing, looking straight ahead. Half the time he didn't even turn the engine off while he waited for Butch. It was smart to stay away when he was like that. But one thing that put Mr. G. in a good mood was a score. And when one of his guys scored big on diamonds—then he was riding high. He just loved having stones around. And the bigger the better. He and Butch were always carrying a bag of rocks somewhere. Usually they'd have them in a brown paper bag. You'd see that bag lying on the front seat of the car, looking like a ham sandwich. Then Mr. G. would smile real big and open it up and show you half a million in loose diamonds.

The Outfit had its own burglary ring—some of the city's most respected cops were in on the deal—and they had pawnshops and fences lined up all over the country, from Chicago to New York and L.A. That meant they had plenty of opportunities to score jewelry, especially diamonds. They almost always took the stones out of the mounts so the fences could turn them without a problem. But Mr. G. always looked them over first and kept a few for himself. He was like a kid in a candy store when it came to diamonds.

Of course, to say Sam Giancana acted like a kid—when it came to anything—would be a pretty good stretch. The guy was about as serious as death and taxes. He grew up poor, on the mean side of town. There was no telling how many guys he'd whacked by the time I met him. He was from the old school all right, but more important, at least when it came to our relationship, Sam Giancana was from a different generation.

You have to remember that this was the sixties. Peace, love, and so on. Not that I was a hippie. I was actually pretty conservative for my generation. But still, there was a generation gap, a big one, between me and Mr. G. I was a rocker, listening to the Beatles. And Mr. G. was a Tony Bennett, "I Left My Heart in San Francisco" kind of guy. His deal was Cadillacs. I was into race cars. You had to figure that we'd clash eventually.

Mr. G. told me a million times that all the noise we made working on race cars would hurt business at the station. He hated the entire race car culture, which I couldn't understand. After all, we're talking about someone who started out as a wheelman. Sam Giancana was a hell of a driver. But for some reason he put auto racing in the same category as rock and roll. And the music of the sixties was not his deal.

One day Mr. G. showed up and it was so loud in the bays that nobody even heard him drive up to the pumps. He just sat there in his car waiting, getting madder and madder, and still nobody came out to pump his gas. All of a sudden I looked up and there he was, standing right behind me. I could tell he was pissed. Right away I motioned to the guys to turn off the engines. It got real quiet in those bays.

Mr. G. had a way of talking when he was mad—it was like a growl. He said, "What the fuck's so important that you can't take care of your customers? Do I get some gas or not?" I said, "Whoa. Hold on a minute." But he was so hot he wasn't hearing a word I said. He was already headed back to his car. I followed him and started pumping his gas, but now he was yelling. He said, "Are you fuckin' crazy? That race car shit's for kids. You wanna be stuck doing that crap for the rest of your life? What the hell are you gonna do with yourself?"

I didn't have an answer. I knew he was right—sure, I was making good money, but where was it all heading? I just looked at him and shook my head. And right then, his whole expression changed. He put his hand on my shoulder and said, "Listen, kid, I'm gonna send a guy around to talk to you. His name's Vinnie Inserro. Maybe you and him can work something out. Who knows, maybe you'll end up with a whole new career."

I'll never forget that. You can say what you want about Sam Giancana. But he cared about me. And at that moment that was all that mattered.

⊕ ⊕ ⊕

You don't all of sudden start having coffee and doughnuts with killers. It's not like you go out of your way to fraternize with that class of people. It happens gradually. First you get acquainted with the bookies, and then here come their buddies the juice guys. Next thing you know, the machine guys are dropping by. Then it's the hustlers like Mugsy Tortorella, with their trucks full of stolen merchandise. They treat you good, put a little extra green in your pocket, and you figure what the hell, who are they hurting anyway? You never see the guys behind it all—the scary sons of bitches like Mad Dog DeStefano. You never even hear their names.

Pretty soon there's a few new faces coming around. You can tell right off these guys are different from the others you've been dealing with. They've got their nice clothes and fancy cars. They've got some class. You're impressed. You figure you're moving up in the world. Eventually they offer you a few bucks for stashing their car, something like that. Nothing illegal. You're just doing them a favor. That's what you think anyway. But right there, that's when you cross the line, and you don't even know it. Not yet anyway.

⊕ ⊕ ⊕

"How's twenty-five hundred bucks?" Vinnie Inserro asked. He handed me an envelope stuffed with cash.

Not bad, I thought—not for closing down my Sunoco on Saturdays so he and his pals could run some high-stakes poker in my back room. I was already closed on Sundays for a craps game Turk and Joe had going. They were giving me fifteen hundred a shot for that little favor. Inserro was offering me almost twice that much to close on Saturdays. That's four grand a week—*not to work*. What was there to think about?

Doing the deal on Sunday with Turk and Joe had been a no-brainer—we were friends. I'd only heard about Inserro—I didn't know the guy, not personally anyway. But Mr. G. had sent him around so that was that. Besides, everybody was high on Inserro. The way Turk and Joe talked, it was like they were related. It looked like working with the guy was going to give me a real boost. I figured how could I go wrong doing the deal?

As it turned out, Pete Altieri had the answer to that question. One night I stopped by Lucky's for a burger. Pete and I started talking, and naturally the situation with Inserro came up. Pete asked if I was going to go in on it, and I told him I couldn't think of any reason why I shouldn't, that it looked like a hell of an opportunity to me.

To my surprise, Pete wasn't at all comfortable with the idea. He said, "This deal with Vinnie's a great opportunity all right . . . 'specially for a guy in your spot . . . a guy just trying to make his way up. But there's a lot at stake with those fuckin' games Inserro's talkin' about. There's times when there's as much as sixty grand on the fuckin' table . . . hell, maybe more. And let me tell you, if there's ever a problem—you know, if somebody's got a beef, hey, whatever—well, then you got everybody up the fucking line on your ass. And it won't be Vinnie Inserro they're gonna look at to make things right, either. Fuck no. It'll be *you*."

Pete didn't say anything else about it, but he didn't have to. Shit always runs downstream. I knew that. And in that particular situation, I was going to be downstream. Still, I wasn't put off. I believed Inserro's deal was a terrific chance for me to get next to some of the big shots—just like Mugsy said. For all I knew, it might be my only chance.

But even more important was the fact that Mr. G. had sent Vinnie Inserro over with the deal. I didn't think it would be wise to back out, not with him involved. But I admit I thought about backing out, especially after my conversation with Pete. From what I knew about Vinnie Inserro, the card game operation was just a sideline. His real deal was murder. Vincent "The Saint" Inserro had whacked dozens of guys, most of them connected to loan-sharking. Lately he was doing hits with Sam Giancana's favorite triggermen: Chuckie Nicoletti, "Milwaukee Phil" Alderisio, and Tony Spilotro. The nice guy who offered me twenty-five hundred bucks to close my joint on Saturdays was a stone killer.

But Vinnie Inserro didn't look like such a bad guy. He didn't act like one, either. He didn't throw his weight around, nothing like that. In fact, we hit it off right away. And as far as threatening was concerned, Vinnie Inserro was at least a dozen years older than I was and a real shrimp—maybe five feet tall and a hundred twenty pounds. But I realized that appearances were misleading. So I was two feet taller and a hundred pounds heavier? Everybody said that if you put a gun or a knife in Vinnie Inserro's hand, he'd make up for his "shortcomings" real fast.

When I got into the gas station business, rubbing elbows with killers wasn't exactly on my list of goals. But making money was. And the bottom line with the Outfit is this: If you want to keep making the green, you don't get shy about having friends with unusual occupations. You go along with the program.

So Vinnie Inserro had a bad rep—and so did his friends. So what? I knew that if I wanted to do this deal, it was going to be with him. This was business, after all. It wasn't like I was inviting a killer into my living room, now was it?

⊕ ⊕ ⊕

The back room at the gas station was no Taj Mahal. It was always cold, which might've been a nice feature in the middle of July, but in the winter, you froze your nuts off. Besides being cold, the room was also real small. Before Turk and Joe started holding their craps games back there, you could hardly walk for all the junk. I had a few old metal chairs we used for sit-downs, but that was it. It was strictly no frills.

Things did improve some after Inserro got involved. We put in a few coolers for beer and a bucket of sand for cigarette butts. I made sure there was toilet paper in the men's room. But even so, that back room was pretty bad. It was still all crapped up. Boxes of car parts were stacked all the way to the ceiling. Tires were rolled over in the corners. Oil cans and carburetors and anything else that didn't have a box were all crammed together on a bunch of shelves. But what can I say? It was a storage room, not the Desert Inn.

None of that stopped Inserro. Given the mess in that back room, he had those games very organized. He had a dealer, a retired guy out of Vegas, come in to handle the table. He had juice guys and a cash box. Sometimes he even had beer and sandwiches. And after a couple of months the games were really taking off. It was all high rollers Inserro was attracting, too. Tons of them. By noon on Saturday there were guys in three-piece suits standing in line outside the station just dying to get in. Pretty soon they got wise and started coming around on Thursdays, just to make sure they'd get in on the next game.

You might wonder why those type guys would want to roll dice or play poker in a joint like my back room. It all boiled down to one thing—money. That's your answer right there. Take the craps games, for instance. Once they got going, there'd be stacks and stacks of hundred-dollar bills lined up all over the floor. To a gambler, the thrill of winning that much dough is irresistible.

Week after week I watched those executives in their pinstriped suits get down on their hands and knees just to roll a pair of dice. Right then I realized how seductive gambling was, how for some guys that's all there is in the world—which was just how it was at the station. It got so that everybody was either a bookie or placing a bet with one.

Most of the Outfit guys loved to gamble. It didn't matter on what, either. Some of them were real obsessed. There were days you would've thought I was running a wire service (a special telephone line that provides gambling information to book joints) instead of a gas station. Thank God gambling didn't do a thing for me. I never even thought about it until I went to work for Pete. But at the gas station you couldn't get away from it. Some guy was always trying to take you to the track or laying odds on one thing or another. If I'd been a gambler, it would've been a constant temptation. But I wasn't interested in throwing money away. I was interested in making it. And I was doing that all right. I was raking in four grand every weekend thanks to those games. Talk about easy street.

Every so often I had to remind myself that it wouldn't last forever, that the weekend money was gravy, not the meat and potatoes. I'd seen it before. One day those guys would be flying high with their games, and then all of sudden it was over. At Tony Pappas's station it had been like that. They had games that went like gangbusters for six weeks and then, *bam,* they'd just burn out. Things would get real quiet, and Tony would piss and moan about missing all that extra dough. But after a few months the guys would drum up another game, and Tony would be smiling again. He always made out real good—he just couldn't count on it.

I was smart enough not to count on it—but for some reason, there never was any slowdown in the action at my station. In fact, things were going so good that Inserro started more games in the other suburbs. At one point he had as many as thirty going at once.

My Sunoco was hitting on all cylinders. Besides those executives I was getting a whole other group of Outfit guys coming by for the games. At first I had no idea who the hell they were, but the respect that was shown by the other guys was all I needed to know. When everybody comes to attention, you know the score. This guy you're seeing doesn't need a name.

Turk and Joe used to kid me around. Turk would say, "If the feds wanna know who's who in the fuckin' syndicate, all they gotta do is come by this joint." At the time, I was proud of that fact. It meant the guys trusted me, that I ran a good operation. And it was true, too—just about everybody that was anybody made it by that Sunoco at one time or another during the time I owned it. There were the regulars, guys who had actually become my friends, like Turk and Joe and Mugsy and their pals. And there were the guys who just wanted to use the phone or maybe drop off a car—which was where Marshall Caifano and Butch Blasi fit in the picture. And there were some of the bigger fish, too—guys the papers called "overlords" because they had their own Outfit territory. They might stop by every now and then for a tank of gas or a Coke, or maybe they were curious about what the deal was out in Summit. This group included men like Gussie Alex, a good-looking Greek the FBI tagged as the "syndicate boss of Chicago's Loop," and Ross Prio, the head of all the far northwest operations.

There were also a few of Sam Giancana's old 42 Gang buddies: Fifi Bucieri, a real sweet-looking, scholarly old guy who handled Outfit business on the West Side and would kill you in a second; Willie "Potatoes" Daddano, a torture nut who controlled all the machines in DuPage County; and Jackie "The Lackie" Cerone, a real tough guy who I'd heard was in on the gruesome Action Jackson murder. Occasionally a few of the Outfit Jews would drop by— the gambling wizards like Lou Lederer and Les "Killer Kane" Kruse, who Turk said were heavy into the Vegas scene as well as gambling rackets overseas.

Naturally, when you've got that many guys coming into your joint, you're not going to hit it off with everybody. That was the case with Vinnie Inserro's pals Chuckie Nicoletti, "Milwaukee Phil" Alderisio, and Tony Spilotro. They were enforcers. I particularly didn't care for Tony Spilotro, who I'd run into back when I worked for Tony Pappas. He was cocky, the type of guy that just loved to bust a guy's balls. He was always pissed off, always looking to nail somebody. "Look at Tony cross-eyed and he'll whack you," Turk had warned me early on. I treated Tony Spilotro with respect whenever he came into the station—which fortunately was rare—but I was always wishing he'd leave. You could've cut the tension with a knife when he was around. With Spilotro dropping by my Sunoco, I figured it was just a matter of time and Mad Dog DeStefano would be walking in the door—something I wasn't exactly hoping for.

⊕ ⊕ ⊕

Inserro's Saturday night games hadn't just affected the weekend crowd, they'd had a big effect on my clientele overall. The games were attracting the big fish, which was great, except for one small detail—as the names got bigger, the stakes got higher. About that time, I started to think more and more about what Pete had said. It might be Inserro's games the guys were coming to, but it was my station. Which meant I had to be on my toes every second. Things had better go nice and smooth. No beefs. No bullshit. Any trouble and I was screwed.

CHAPTER 11

In the Outfit, it's the bad situations that you never see coming. Things can be going along just fine, but then overnight, you're screwed. One morning a guy's your best friend, by noon he's digging the hole where he and the rest of your pals are going to plant you that night. You may think there's something between the two of you, but really you're nothing to him.

I never met a guy who saw that right off. Then again, who would want to? Who wants to admit he's just another cog in the machine? And that's what the Outfit is—a machine that's well oiled, perfectly tuned, and every part dispensable. That includes your best buddy. And those guys who are going to try to whack you someday. It even includes the "boss of bosses." So why should you be any different? You're a nobody in the deal. If it gives you any comfort, everybody's a nobody in the Outfit. The Outfit doesn't need you or anybody else to survive. The Outfit has a life of its own.

If you're lucky, one day you realize all that and you get a wake-up call. It's like losing your virginity or killing a guy. After that, you're never the same—not once you know the score. And that's just how it happened with me, too. Even now, when I stop and think about it, it's like it was just yesterday. But it wasn't. It was thirty years ago. And yeah, things were different after that. Just not as different as I thought.

⊕ ⊕ ⊕

It was early on a Saturday afternoon when Inserro got a call from Chuckie Nicoletti. Nicoletti said he and Phil Alderisio were going to bring over this Outfit big shot named Murray Humphreys for the games that night. Normally not much fazed Vinnie Inserro, but this time it wasn't ten minutes after he got Nicoletti's call and he was acting squirrelly.

Even though the station was closed on Saturdays, I usually came in. I liked to have some time alone, just to fiddle around on a car or two at my leisure—which was what I'd been doing when Inserro got the news that Murray Humphreys was coming over that night. For a while, I just kept right on working. I pretended not to notice how Inserro was acting. But after an hour or so, I figured he needed to lighten up; I handed him a beer and waved him into the back room.

I expected Vinnie to mellow out after a beer or two—not give me a history lesson on Murray Humphreys. From what Inserro said, Humphreys was in the very top echelon of the Chicago Outfit and had been around since the days of Capone. When Scarface was sent up, the Chicago Crime Commission named Humphreys Public Enemy Number One, meaning he was next in line on the feds' wish list. But Humphreys didn't make boss, Frank Nitti did instead. And after that, Humphreys got into slapping backs and greasing palms. Evidently he was pretty good at it, too—because Murray Humphreys was considered one of the biggest political fixers in the history of organized crime. His influence spread as far west as California and as far south as Cuba and Guatemala.

According to Inserro, Humphreys had even helped Bugsy Siegel and Meyer Lansky put together the first hotel in Las Vegas, the Flamingo. And he'd made sure Chicago kept involved out there by lining up a real sweet deal with Red Dorfman and the Teamsters Union—which explained how the union's fat pension fund got invested in the Las Vegas casinos in the first place. Humphreys also collaborated with Chicago's Johnny Roselli in Hollywood, where, Inserro said, the Outfit kept a firm grip on the studios.

By the time the late fifties rolled around, it was pretty well understood that Murray Humphreys had dined with presidents and kings all over the world, from the Philippines to Iran. But his real claim to fame was the fact that he'd single-handedly put some of the nation's richest labor unions in Chicago's pocket—a move that was worth billions to the Outfit, particularly in the gambling industry.

With a résumé like that, you could understand why Sam Giancana made Murray Humphreys one of his top financial advisers. But it wasn't just his brains that attracted Mr. G.; Humphreys had class, too. He could fit in with the country club set as well as politicians and executives. He was also about as cold-blooded as a guy could get, which was a real interesting combination for a guy in his position. It made him even more dangerous. Like Inserro said, "Murray Humphreys is a real gentleman. But he's tough as fuckin' nails, too. And another thing—Murray Humphreys and Sam Giancana . . ." Inserro crossed his fingers. "They're like this."

Right up to the minute Humphreys walked into the back room with Alderisio and Nicoletti, Inserro was all over the place. If he checked the coolers once to make sure there was enough ice, he checked them a hundred times. He put up a new card table. He brought in some fancy mixed nuts and cases of good booze. He even had the place swept up.

I figured if a stone-cold guy like Vinnie Inserro was that worked up over a deal, then how should I feel? It was *my* joint the guy was coming to. Most times I didn't stick around for Inserro's games—but no way was I going anywhere that night. That night the stakes were just too high. And I wasn't so sure the odds were stacked in my favor.

⊕ ⊕ ⊕

By midnight there were over a dozen, maybe two dozen guys in the back room. It was elbow to elbow, standing room only. Guys were leaning against walls, sitting on boxes. There were six for poker: Nicoletti, Alderisio, and their "guest," Murray Humphreys—plus

Needles Gianola, who was one of Mugsy's longtime sidekicks, and my old pals Turk Torello and Joe Ferriola. Between them, they'd probably made more than one funeral director a very rich man. Naturally, Vinnie Inserro had every intention of keeping them happy. He'd even gone so far as to hire a bartender and cocktail waitress and a couple of broads to hand out sandwiches and Cuban cigars.

Pretty soon the back room started to get a little atmosphere. The smoke was so thick you could barely see. We had Sinatra playing on a record player. The booze was flowing real nice. The poker table was covered up in dough, stacks and stacks of it, six inches deep. We're talking some serious money. But it could have been ten bucks the way those guys were acting. By that time they were all getting loaded. Even Inserro was starting to relax.

And that's just when it happened—these three big bruiser type guys barged right through the door. They were waving .38s, yelling at everybody to get their hands up. They didn't even wear masks. They waltzed in like they owned the place. I thought they were either the dumbest guys in Chicago or they had the biggest balls I'd ever seen.

I just knew Vinnie Inserro had to be shitting his pants. I sure as hell was. I knew that if we'd had a few guys outside watching the joint this wouldn't have happened. And there was another aspect of the situation I wasn't too happy about, either. There was something about these stickup guys that looked familiar. It was killing me, trying to figure out where our paths had crossed. I definitely didn't want anybody getting the idea we were acquainted, because somebody was going to pay big-time for this deal, and I didn't want it to be me.

I looked around the back room and my stomach turned. Sure, all the guys were at a disadvantage now—they were at gunpoint, with their hands in the air—but we're talking a room full of hitters. These three guys were holding up men who killed people for a living. And at that particular moment, they were all dying to kill a few more.

You could've heard a pin drop. Nobody said a word while we emptied our pockets. Those bums took our watches, our rings—everything that was worth two bits, they stole. They must have taken a half dozen guns off us before they were done, too, which was one of the few intelligent things they did.

When they were done taking all our loot and guns, they raked the cash off the table. Then those lousy sons of bitches left the same way they came in: they walked.

⊕ ⊕ ⊕

I didn't want to think about what had happened at my Sunoco. Needless to say, I was very concerned. Even more so when I learned that the guys had made off with almost a hundred grand. But when I got home that morning and opened the Sunday paper and turned to the sports section, I really started to sweat. It suddenly came to me where I'd seen those stickup guys—they were the Bastone brothers, from my old racing days. I started wondering what would happen if one of the guys found out I knew them; after all, the Bastones weren't exactly unknowns. It was like my worst nightmare, having something like that happen at my station.

And when Turk Torello called, I knew I had a serious problem; the guys were thinking the holdup had been an inside job. When Turk started firing questions at me about it, I figured they were on to something, that somehow they'd gotten the idea I was involved in the deal, that maybe they already knew about the Bastones. I knew I had to play this deal very carefully. I told Turk it looked like an inside job to me, too—but that even so, I couldn't imagine what guy would be so stupid as to set us all up like that. I said the word *us* in such a way that I made sure Turk understood that I wasn't a rat, that I was one of them.

"Yeah," Turk agreed. "Any guy involved in a deal like that holdup has gotta be one dumb son of a bitch all right."

I didn't have anything to say to that. I knew the reality of the situation. Turk had made his point. I was in the jackpot. His call was meant as a warning—maybe even a threat.

⊕ ⊕ ⊕

Later that afternoon Joe Ferriola dropped by the house—which was something he'd never done. I invited him in, but he said he was in a hurry, that he'd just stopped by to see if I'd come into the station later. I told him yeah, sure, I'd be over, no problem. I said I'd bring over a pizza, maybe some beer, that we'd shoot the breeze. And he just said yeah, yeah, yeah, like whatever I said was good with him. Then he went on his way. I could tell it was all bullshit. I knew this was about the holdup.

All the way to the station that night, I was wondering if this was it for me—the end of the line for Mike Corbitt. I'd been around long enough that I'd heard the stories, that it wasn't your enemy who took you down, it was always your best friend. But I still hadn't reached the point where I actually believed it. I still thought I was special to those guys. Joe and Turk were probably the closest to being my friends, aside from Pete Altieri. I couldn't imagine Pete and those guys taking me out. Not them. Not me.

You can imagine my surprise when I drove up to the station and there was Pete standing in the doorway, with Joe and Turk right beside him. I'd never seen Pete at one of those Sunday night games. When Pete smiled and waved me in, my heart dropped to my knees.

Inside the station it was real quiet. But it shouldn't have been quiet. There should've been a few guys hanging around. I was starting to sweat pretty good now. But I didn't show my hand. I pretended things were just fine. I handed Pete the pizza and carried the beer inside.

I was hardly in the door and they were trying to make me comfortable. I figured they were hoping I'd let down my guard. Pete said that Turk and Joe had called off the games that night on account of the holdup. He said Inserro was waiting in the back room and wanted to talk about the stickup and how we could avoid a situation like that happening again. But I wasn't buying it. I didn't trust Joe and Turk. I didn't even trust Pete. And now, hearing that Vinnie Inserro was waiting for me in the back room—well, I figured I was screwed. My wheels were turning. I was thinking about how there was a door leading out of the back room into the lot behind the station. There weren't any lights out back. It would be pitch-black. I figured one of them was going to pop me, and that would be the end of it. They'd haul my body out and throw it in a trunk. Later they'd dump me in one of the canals or maybe the forest preserve.

As you can imagine, I wasn't happy about walking down that hall to the back room. They let me lead the way, too—which meant I had my back to them the whole time. It was a long walk, like I was going to a execution, but I kept up my front; you would've thought we were going to a picnic the way I acted. Inserro stood up when we walked in. He was very friendly. Too friendly, I thought. We sat down and Turk passed the beer around. It was all very social. And that really bothered me. Inserro was not at all worked up, and I thought he should have been. Instead the guy was all smiles.

Inserro started out by saying that the reason he wanted to talk to me was because some people were saying the holdup was an inside job. Word had it, he said, that there was a rat working at my station. If that was the case, then he figured I might have some idea who it was.

The guys started asking me all about who I had working for me. They wanted to know what I'd said to Annette about the games. They even asked about my folks, my kid brother, my sisters. When they brought my family into the deal, that got to me.

I told the guys in no uncertain terms that I'd never talked business around any of my family, and as far as my mechanics were concerned, they might not all be Italians, but they knew what the word *omerta* meant. They kept their noses clean.

Everybody was smiling and nodding, like they totally believed every word I was saying. Turk said it was good to hear my side of things because now they'd look into other possibilities. He said they knew I'd be concerned about the holdup and that I was probably thinking there might be some repercussions for me personally, being it was my joint where the deal came down. Inserro added that they wanted me to know that no matter what anybody else was saying, none of them bought it. "We know there's no way in hell you'd set us up," Pete chimed in. "Hell, we're your friends. Right?" He reached over and slapped me on the back, and right then my stomach turned. I couldn't believe my ears. *Friends*. It was just like the old-timers said. It was always your friends who put you in the ground. It hadn't been twenty-four hours and I was already being fingered as the guy who gave everybody up.

I should have been scared, but for some reason I was mad as hell. I gave it back to them real good. I told them they were right, there was no way I'd ever do something like that. But forget all that crap about being friends. The fact was, I wasn't an idiot—friends didn't have a thing to do with it. Immediately they all started backpedaling. Inserro said he knew I was a stand-up guy. He said they'd find out who the stickup guys were, and when they did, he'd let me know. And that was the end of it. We all shook hands and walked out.

But that wasn't the end of it for me. For days I was one paranoid son of a bitch. I was looking over my shoulder all the time. Every hour was like an eternity. I was thinking about those goddamned Bastone brothers and wondering when the hammer was going to fall.

A week had gone by when I got a call from Pete, asking me to come by Lucky's after closing that night. He said he had some news about the holdup. I was feeling real low after Pete's call. Here's one of the few guys in the world I figured gave a fuck about me, and what do you know—he's the one who was probably going to take me out.

That night I went over to Lucky's after closing. The place was pretty dark. We sat down in a booth and Pete poured me a cup of coffee. I was trying to hide my nervousness. I was looking around, sizing things up. The joint looked like it was empty, but an entire platoon could have been hiding out in the kitchen. You'd never know. Not until it was too late. Needless to say, I didn't like the situation.

Pete didn't get down to business right away. Instead he started talking about the old days, back when I was a kid: the time we first met, all the fights I got in, stuff about when I worked at A&W. Pete's little walk down memory lane wasn't making me feel any better. It was like I was already dead and the guy was doing my eulogy. Finally I couldn't stand it anymore and said, "Listen, I didn't come over here at one o'clock in the morning to talk about this crap. You said you had some information about the holdup . . . so what the hell is it?"

Pete looked me in the eye and said, "You ever hear of the Bastones?"

It was like a punch in the gut, but I tried to act very nonchalant. "The Bastones? Yeah, sure. The race car guys. Carmen, Sal, Angelo . . ."

"Yeah, that's them." Pete's eyes narrowed and he looked me over. He knew he had me, and I knew he knew. "So how come you didn't say it was them that did that holdup?" he demanded.

I tried to look surprised, like the fact that the Bastones were the stickup guys was a total revelation to me. I told him I found it hard to believe the Bastones were into stickups. I said I'd never thought they were geniuses, but I didn't think they were crazy enough to try a stunt like that. I added that it had been years since I'd been around them, which explained why I wouldn't have recognized them at the station.

I could tell Pete didn't buy it. But he didn't say that. In fact, he didn't say a word. He just looked at me. Pete Altieri had never looked at me like that. It was like I was a total stranger. A light went off in my head and I realized that when push came to shove I was nobody to Pete Altieri. I was nobody to any of those guys. All that mattered was the organization. The organization was bigger and more important than anybody or anything. Bigger than me or Pete or anybody else I might ever care about.

Pete smashed out his cigarette and smiled. "As far as the guys are concerned," he said, "you're in the clear on the deal. You hear what I'm saying? Forget about it." And that's when I knew it was really over; I was off the hook.

⊕ ⊕ ⊕

The Bastones were off the hook, too. There were no repercussions for them whatsoever. Some other guys might have tried to pull a stunt like that, and *bam*, it would have been all over. Later I'd find out that they were related to Joe Ferriola; that's what saved them. Plus the fact that they returned the money.

It would be ten years before the Bastones crossed my path again. Then I'd realize that Sal and Carmen Bastone hit the lottery that night at my station. At the time I just figured they were some of the luckiest bastards I ever met. But someday the name Bastone would be known throughout the underworld. And by that time, my fate— and quite a bit of the free world's—would rest in their hands.

CHAPTER 12

Sam Giancana was starting to drive away when he stopped and rolled down the window. "You know," he said, like it was an afterthought, "this gas station is goin'."

I stopped in my tracks. "What do you mean, *going*?" My heart began to pound.

"We're gonna tear it down, put in a parking lot."

"Nobody told me about that."

Mr. G. looked amused. "Well, I guess they told you now . . . didn't they?"

He gave me a wave and drove off. Tearing down my station was no big deal to him. But that Sunoco was all I had. It was my life.

⊕ ⊕ ⊕

One of the things that gives the Outfit its power is its unpredictability. One day a guy might be planted for mouthing off to the friend of a friend of some guy he's never even met. Another day he could go right up and call the same guy a fucking rat bastard and all he'll get is a laugh. Take the deal with the Bastones. They should have gotten whacked; instead they got a boost and started working for Mad Dog DeStefano as juice collectors.

At the time, Sam Giancana was the most unpredictable of all the men in the Outfit. "It keeps the guys on their toes," he used to say. But it was more than that. The truth is, if you keep a guy always wondering, then you've got him afraid. And if he's afraid, he's all yours.

I was just sick, knowing that my gas station was coming down. I could hardly work after Mr. G. came by and dropped that bombshell. I was wondering what I was going to tell my folks. And what about my wife? We had a baby now, little Keith, what about him? None of the guys gave a rat's ass about my family. They had no idea what I was going to do to make a living. And they didn't care, either. But working at that station was all I knew. I'd never considered that it might all come to an end.

That night I broke the news to Annette. She was standing in the kitchen, washing dishes. When I told her, she leaned over the sink and started to cry. I felt so bad. Finally she wiped her eyes with the dish towel and turned around to face me. "You know we have a son now," she said. "What are you gonna do? How can you put food on the table without a job? How can you pay our bills?" Her voice cracked and she broke into sobs. "How, Mike . . . *how*?" What could I say? I didn't have an answer. I didn't have the slightest idea how we were going to make it.

⊕ ⊕ ⊕

I was in a fog. I couldn't sleep. I couldn't eat. All I could hear were Annette's words ringing in ears, "How, Mike, how?" A week passed and I still didn't have an answer to that question when Marshall Caifano came by and told me Mr. G. wanted me to know I shouldn't worry about anything. "Relax," Marshall said. "You'll have a spot."

Relax? Was he nuts? I had no idea when they were going to close the station and tear it down. Plus, I had no opportunities—at least any that I knew about. Nobody was telling me anything. I was just going about my job, waiting. Mr. G. would come by on Fridays and leave his car, and I'd just look at him, like—okay, so where's this job I've been hearing about? And he'd grin and say, "Relax, kid. Don't worry about it. We'll have something for you real soon."

It wasn't long before Annette started pressuring me to get another job. She had no idea what I was involved in or who I was involved with. And I didn't tell her, either. Besides, there was no way she'd understand that if Mr. G. told you to *wait*—that's what you had to do.

Weeks went by. Annette and I were having some major difficulties, and I was very low, but not as low as on the day Mr. G. came into the station and laid his next surprise on me. Before I could even offer him a chair, he said, "Hey kid, how would you like to be a cop?"

I couldn't believe my ears. "You're kidding me, right?"

"Hell no. We got the mayor out in Willow Springs in our pocket. He's our guy. He tells me you can be a cop in one day."

I was ready to lose it. I couldn't believe that this was what I'd been waiting for—this was supposed to be my big opportunity? Me, being a cop? I forgot myself. "Are you crazy?" I said. "You want me to go out to Willow Springs and be a cop? Are you kiddin'? No fuckin' way."

Mr. G. started to laugh. "Hey, calm down, will you? What's the big deal about being a cop? I was thinking that maybe you could take care of us out there. If it's the money you're worried about, well . . . I don't know exactly what you'll make. It won't be as much as here at the station . . . at first anyway. But it'll work out." He could tell I wasn't convinced. "Give it some thought, okay? You don't have to give me your answer right now. I'll come by in a few days. For now, you take some time and think about it. Not too long, though. The station's comin' down next week."

My heart sank. I still couldn't believe it was being torn down. Naturally, I told him I'd think it over. But I was in shock. I was pissed, too.

A few days later Mr. G. came by again. He wanted an answer. He said, "So what's it gonna be, kid? You wanna be a cop?" He was looking at me real good. His eyes were going right through me.

I had absolutely no desire to be a cop. But what did that matter? What choice did I have? Was I going to tell Sam Giancana, the boss of bosses, no? In the Outfit, if you're smart, you go with the flow.

Nothing's ever for sure. You'd better get it while you can—before it gets you. So I looked Mr. G. right in the eye and I said, "Yeah, sure. A cop. I'll give it a try."

"Good," he said. "But just remember, kid . . . don't forget who your friends are." And then Sam Giancana winked, like me and him had this big secret, like now I was on the inside. And all of a sudden I knew I'd made the right decision. Yeah, I was going to be a cop.

⊕ ⊕ ⊕

To be a cop in Willow Springs, you had to be sworn in by the mayor, a guy named John "Doc" Rust. As it turned out, I already knew Doc Rust—or should I say, knew *of* him. Our paths had crossed before, when I was just a kid working at A&W, picking up slots and pinball machines at one of his joints, the White Star Inn. From what I'd heard about Doc Rust, the title of mayor was a pretty loose term. Doc Rust was just your common everyday drunken bum, as well as a devious crook, a con man, and richer than God. He'd also won every election he was ever in by a landslide. The whole town loved him.

The Outfit had its own love affair going with Doc Rust. He was the man Mr. G. referred to as "their guy" in Willow Springs. And from what I was told, Doc had been their guy since 1935, when he'd first made mayor. Doc's father, who really *was* a medical doctor as well as the village mayor, had also been "their guy" during Prohibition.

The last time I'd seen Doc Rust he was all decked out in a beat-up Texas Stetson and a pair of cowboy boots, sitting in the White Star Inn at a corner table with a bottle of Christian Brothers to keep him company. And now here it was almost ten years later, and it was like the guy hadn't moved an inch. It was only nine in the morning and he was already sitting at that corner table, wearing those same clothes, sucking on a bottle of Christian Brothers.

Right away I could tell Doc was plastered. And I figured there was no way I was going to be sworn in by a guy in his condition. At the time, I had no idea that being drunk was a way of life for Doc

Rust and a little booze never stopped him from conducting village business.

When Doc offered me a drink, I told him no thanks, but he gave me this real stern look and told me to be a "good laddie" and accept his hospitality. There was no use arguing with the old guy about it, so I told him sure, pour me a drink, and I pulled up a chair. Right away I realized that to get along with Doc, all I had to do was listen and nod constantly. So how hard could that be? It was gut-wrenching hard, that's what it was, listening to him go on and on about his "wonderful little village" and the "sleepy little town" of Willow Springs. I smiled and nodded at Doc Rust until my jaws ached and my neck went stiff.

But the worst part about it was putting up with the disgusting White Owl cigar he had clenched between his teeth. That was the real test. And was I ever failing. That cigar was killing me. The smoke was like toxic fumes. My eyes started watering. And the stench was just tremendous, really nauseating. Today a guy would go to jail for smoking something like that.

One thing about Doc—the guy had no class whatsoever. He didn't know the meaning of the word *sip*. He guzzled a drink down, smacked his lips, and went right back to talking—and pouring his next one. After a couple of hours of Doc "wetting his whistle" like that—and going through God knows how many bottles of brandy—I was starting to wonder if we'd ever get out of that tavern and get down to the business of making me an official Willow Springs cop. But when I dropped a hint that maybe we should head over to his office, he looked at me like I was nuts. He said he didn't need an office. An office, he informed me, was nothing but a waste of the taxpayers' hard-earned money.

When Doc finished his speech about how much money he'd saved the citizens of Willow Springs, he calmed down. He said he'd had an office once, a medical office, back when he was "practicing medicine"—a claim that struck me as peculiar since I happened to know he wasn't a real doctor. Being called "Doc" made him feel important—like being mayor. But it was all a sham. Then again, what was a guy like me doing in law enforcement?

Doc swore me in that afternoon, right there at that table in the White Star Inn. It was a very momentous occasion. We were already sloshed when he called for our third bottle of Christian Brothers and made a toast in my honor. "A policeman's toast," he said and raised his glass. "May all your roads be paved with silver and your pockets lined with gold."

At the time, I had no idea what the old man was talking about. I was convinced he was crazy as a loon. But it didn't take me long to figure out that Doc Rust was on to something out there in Willow Springs. Silver and gold? We're talking greenbacks here. Millions and millions of greenbacks.

⊕ ⊕ ⊕

When I reported for work at the Willow Springs Village Hall the next day, I was in for a real surprise—but not half as big as the one Chief Joe Kresser got when he saw me standing on his doorstep. He didn't even know I'd been hired.

Joe Kresser was in his mid-sixties by the time I met him. He and Doc went all the back to Capone and the bootlegging era. Kresser's office at the police station was just about as impressive as Doc's was at the White Star Inn—only worse. It was nothing but a miserable one-room shack. But the thing that really got my attention was how quiet it was, like a funeral parlor. At my gas station, things had always been hopping. The phones were constantly ringing off the wall. Guys were coming and going. There was action all the time. But this joint of Kresser's was dead. It was also very old-fashioned, nowhere close to being a modern law enforcement office.

I wasn't thrilled to see where I was going to be working. I was pretty sure a guy couldn't do much crime-fighting under the conditions they had going. I didn't know it at the time, but that was the whole idea. Enforcing the law was not the first priority of Chief Kresser and Mayor Rust.

When Kresser saw me standing in front of his desk, he didn't exactly throw out the welcome mat. He was reading the newspaper and didn't even look up. He just said, "What the hell do you want?" and turned the page. So right away I was annoyed with the guy, but I was determined to keep my cool. I told him my name and said I was the new policeman, and he nodded and kept right on looking at the paper.

After a few minutes of me just standing there watching him read, I started getting real pissed. But I couldn't just walk out; the job was too important. I decided to give it another try, but before I could get a word out, Kresser said, in this very snide tone of voice, "Go on and get the hell outta here. I got no time for your bullshit." When he said that—well, I almost went nuts. I'd never even considered the possibility that the chief of police might not be expecting me, that this Doc Rust character probably wouldn't even remember swearing me in.

I was sick. I'd been counting on that job. I had a ton of bills to pay. I was asking myself what I was going to do. I was about to come out of my skin. So when Kresser turned the page of that paper again and kept on reading like I wasn't even there—well, that did it. I exploded. I tore the newspaper right out of his hands, leaned over his desk, and said, very calmly, "Look, like I told you before, I'm your new policeman. Doc Rust swore me in yesterday and he told me to come see you this morning. If you don't believe me, then get on the phone and call him. I'll wait."

When Kresser heard Doc's name, his mouth hit the floor. He started stuttering and stammering, "Are you trying to tell me you know Doc Rust? If that's the case, then you tell me . . . where's he at right now?"

I had to smile. I said Doc was over at the White Star Inn. Now Kresser was flustered. He got right on the phone, and when he hung up, he was a changed man. He shook my hand and got right down to business. "First thing we gotta do, son," he said, "is get you downtown and get your uniform."

From then on, it was like me and Chief Kresser were best buddies. That afternoon he drove me down to Kale's, the store in Chicago where all the policemen bought their stuff, and I got my uniform. When we got back to the station, he gave me a badge, a ticket book, and a gun. So now it was official. Mike Corbitt was a cop.

⊕ ⊕ ⊕

You can't imagine the stir my new job as a police officer caused in my family. That night Annette fixed a special dinner, and Mom and Dad came over. Naturally, they wanted to see what I looked like in my uniform. Annette just went wild when she saw me in it, but it was my dad who was the most excited. He had me stand in the middle of the living room floor and turn around again and again. He wanted to see that uniform from every angle. He was so proud of me, he could hardly stand it. After what they'd been through with me as a kid—well, I guess Dad figured my life had completely turned around. He'd been happy things had gone so well over at the Sunoco, but this was even better in his mind. Being a cop meant you were an all-American stars-and-stripes good guy.

It was a great night. I must've looked at my badge and ticket book a hundred times. I was on top of the world. We all were—except Mom. She was a different story. I could tell she was very skeptical. Dad was already in the car, waiting to head for home, when she let me have it. She said I might fool everybody else, but I wasn't going to fool her. She told me she knew all about Willow Springs and all those gangsters out there. I tried to tell her that I was a real cop, that I wouldn't have gotten a uniform and a badge if I wasn't. But she just shook her finger in my face and said, "That uniform don't mean shit to me, Mike. That's nothin' but criminals and thieves out there. You know it and I know it. And that's who you'll be working for, too, criminals and crooks. As far as I'm concerned we can just plan on you goin' off to jail someday. Sooner or later, badge or no badge, that's what it's gonna be. My boy's gonna be in jail."

She started off down the sidewalk toward the car, and I thought that was the end of it, but all of sudden she stopped and turned around. "I don't intend to say another word about this talk we had here tonight," she whispered. "I don't want to hurt your father." For a minute I thought she was going to break down, but then she set her jaw and went on, "Besides, what good would it do? Your dad wouldn't believe it anyway." She gave me a real hard look. "He thinks you hung the moon."

To say that Mom had just rained on my parade was an understatement. Looking back now, I know it wasn't just her attitude about the deal that got me down. I was pretty uncomfortable with being a cop to begin with. I had no idea how things were going to turn out in Willow Springs, and I also knew she was right: this wasn't going to be law enforcement at its finest. One look at Doc Rust and Joe Kresser and you knew that.

It wasn't like I'd hit the jackpot, either. I was only going to be making six thousand bucks a year, plus a one-hundred-fifty-dollar uniform allowance. I had no insurance, no pension, and no benefits. I had no idea where the money would come from. I was already missing that Sunoco. But hey, let's be honest. What I was really missing was all that green.

⊕ ⊕ ⊕

My first night on the job, Chief Kresser put me on with a guy named Art Doogan. Doogan had been a cop for at least twenty years and was an older guy, way over sixty. Kresser said Doogan would teach me the ropes.

Doogan drove us out to Mannheim Road and parked. We didn't have radar then. Nothing like that. We had a timer, a sort of stopwatch, attached to the dashboard. You'd get behind a car and hit a button on the timer and pace the car for a mile or two. That way you'd know how fast the car was going. Too fast and we'd get out the ticket book. Those timers were worthless, total crap. You could manipulate them anytime you wanted to—all we did was get our car up above the speed limit and record our own time. Later, when we wanted to pinch a guy, we'd use the speed we'd recorded earlier to nail him.

That night we pulled over seven or eight cars for speeding violations, and not once did Doogan take out his ticket book. Not once. I just couldn't understand it. I was very proud of my ticket book. I'd spent the whole day making out tickets ahead of time so all I had to do was put a guy's name in, the speed he was going, and all that. I was real disappointed when I didn't get to give out any tickets. I was also bored as hell, sitting in that car with Doogan "teaching me the ropes" all night.

It was sunrise before we finally left our spot on Mannheim and headed to a diner. I waited in the car while Doogan went inside. After a few minutes he came out with coffee and doughnuts. When he got in the car, he handed me an envelope. He didn't say anything, like what the envelope was for. At first I just looked at it. I'd been up all night and I was tired. I wasn't in the mood for practical jokes. Finally I held it up and said, "Okay, so what's the deal?"

Doogan swallowed a hunk of doughnut and said. "You're a real fuckin' choir boy, aren't you? What the hell do you think it is . . . it's your fuckin' end."

"My end of what?"

That did it. Doogan told me to shut up and quit screwing around. I could tell he was very annoyed. I figured if it was that important to him, I'd play along with his little gag and open the damned thing.

There's nothing like the experience of opening your first envelope. Nothing can compare to it, except maybe popping your first cherry. When I saw a hundred and sixty bucks, cash, inside that envelope—well, the feeling was just tremendous. My first night on duty, I'd made almost two hundred dollars. Maybe I'd be able to pay my bills after all. I couldn't wipe the grin off my face. I think Doogan must've realized by this time that this really was all new to me, because he chuckled and shook his head and said, "You'd better get used to it, kid."

We hadn't left that diner and I was already figuring out ways to make more. All day long my wheels were turning. I figured if I worked five nights a week, I could pick up an extra grand above and beyond my so-called salary. In a month, that came to four thousand. In a year, almost fifty. If I worked real hard I could probably do even better. I had to think there were a lot more drivers out there just dying to avoid a speeding ticket.

I admit I never once questioned whether it was right or wrong to do that to people. In Willow Springs, corruption was standard operating procedure. In fact, that sleepy little town had the reputation of being more corrupt than anyplace around Chicago. It wasn't the local citizens who made it one of the nation's most notorious crime capitals. The credit for that went to the tireless entrepreneurial efforts of the police chief and the mayor. After only a few days on the job, I came to the conclusion that Rust and Kresser had to be the most corrupt individuals who ever walked the face of the earth.

Doc and Kresser even shook down Oscar Meyer. Every truck that came through Willow Springs had to stop by the police station. It was understood, like an unwritten law. They *had* to stop. The first time I saw one of those truck drivers pull up in front of the police station and come in with a tray of cold cuts, I said, "What's this?" I was such a rookie. Eventually I got wise to the deal. I was seeing the bread man, the bakery man, the soda man, you name it—they were all pulling up in front of the station and bringing their shit in. It was like they were paying tribute just for being allowed to drive through the village.

Doc and Kresser had a million and one ways to make money. There was nothing sacred as far as they were concerned. Everything was on the block in Willow Springs: broads, booze, even badges. There was always somebody who thought that having a real cop's badge would be the neatest thing since sliced bread and didn't mind shelling out a few bucks for one. Most guys didn't do much of anything with it—Doc said maybe they'd flash it around at a party to impress their pals or a girlfriend. Doc and Kresser got their money and the guy had a little fun. No harm done. And generally speaking, that's the way all their scams went. They were what you'd call "victimless crimes."

Still, for a rookie cop like me, it was Corruption 101 thanks to the Willow Springs on-the-job training program. But that's all I got. When it came to actual police procedure, forget about it. There was no manual, no nothing. They didn't even show me how to use a gun. I had to go over to a gun shop in Summit—on my own time—

and pay out of my own pocket for target practice at an indoor range.

Most guys on the straight and narrow would've quit right there, but not me. I saw the opportunity. From the minute Art Doogan handed me that envelope, I was hooked. I knew I was going to learn by example, from some of the nation's finest.

CHAPTER 13

In 1965 there were only about twenty-five hundred actual residents in Willow Springs, but there were at least forty taverns, ten strip clubs, five houses of prostitution, three casinos, and God knows how many book joints. There were machines everywhere. Places had walls full of slots and rooms full of pinballs. Nobody else in Cook County could get away with that. They didn't even have machines to that extent in Cicero. Short of Las Vegas, there wasn't another spot in America with so many different rackets going full blast, twenty-four hours a day, with no concern whatsoever for the law or the consequences of breaking it. And there I was—this greenhorn rookie cop—right in the middle of it all.

Calling what I did "law enforcement" was a bit of a stretch. Everybody knew my real job was to make damned sure nothing got in the way of all those rackets and Doc Rust's pursuit of the almighty dollar. Being a cop out there gave new meaning to the phrase *law and order.* Most of the crimes didn't have a face—unless it was some sucker's wife and kids. Who cares if a guy pays to get his knob polished? Or an old-timer gets his rocks off playing the ponies? The crimes I saw were about money. Sure, it was money made in ways that society called illegal, but so what? Like Doc said,

"If a guy pays for it, he's a fuckin' criminal. If he gets it for free, he's lucky."

But you could forget about a free ride in Willow Springs. Doc made sure everybody paid his dues. It was like we were running a big corporation with a bunch of small subsidiaries. Those taverns were crappy little joints, but Doc and his Outfit pals operated just like you would in any legitimate business. Everybody's role in the deal was laid out very clearly. Doc was the town boss; he ran everything. Kresser and me and the rest of the police force were what you'd call Doc's bagmen and muscle. Every dime that was made, Doc took his share off the top. From him, it went to a couple of my old gas station buddies: Turk Torello and Joe Ferriola.

Willow Springs was part of Turk and Joe's territory—they controlled *everything* out there. I hadn't realized they'd be around when I agreed to be a cop in Willow Springs. We'd never discussed the particulars of their situation—things like territory and where they stood in the Outfit hierarchy and chain of command. Generally you pick up that type of information over time. Which was the way I figured out how the money from all those rackets in Willow Springs moved through Turk and Joe and on up the ladder.

Like they say, follow the money. In the Outfit you do that, and most times it'll take you right to the top. In this case, Turk and Joe passed the dough over to Cicero and the guys in charge there: Joey Aiuppa and Gus Alex. From them it went right to the top, meaning Sam Giancana. Sam might dole it out to his other buddies, but that was his call. He was the boss.

The Outfit's operation in Willow Springs ran smooth as silk. But it hadn't happened overnight; the way Doc told it, Willow Springs' ties to the Outfit went back to the twenties, when he was running "shine for Capone's boys," and everybody in town called his father "Doc" and him "Little Doc." There wasn't any doubt that Doc had been attracted to all the dough he saw changing hands during Prohibition, but it wasn't just about money. It was what all that dough could do for a guy that hooked him. It was the power. And the control. Doc told me that watching Chicago's early gangs conduct business taught him one of life's most important lessons: No matter what game you play, the odds are always in favor of the house. He said that once he realized that, well—there was no way he was going to be like those saps he saw dropping money in machines or paying whores for a blow job. He was going to be the man in charge.

Looking back now, I imagine Doc's career path turned out way better than he'd ever dreamed it would when he was running booze. His little kingdom might've been nothing but two-bit dives—but that old drunk probably had more money than the Queen of England. Doc Rust was "the house" all right. And he was the boss *and* the mayor. With him in charge, nobody ever had to worry about getting pinched. Talk about power. To a guy in the rackets, having somebody like Doc in a town like Willow Springs was better than dying and going to heaven.

The bookies loved it in Willow Springs. It was amazing the way they conducted business. Gambling was no big deal. All the taverns had blackboards with the results of the day's races from all over the country. There were hydraulics in the walls—left over from the Roaring Twenties—so if there was a raid, the bartender pulled this lever and a fake wall would come down. Not that they ever had a raid in Willow Springs. With Doc in charge, the odds were better they'd have a nuclear blast.

As far as prostitution in Willow Springs was concerned—back then it was right in your face. All the whorehouses were right out in the open. The rest of the clubs in town were just fronts for flesh peddlers, with dancers who were basically glorified hookers. Most of them had these real dingy little rooms in the back, maybe the size of a closet, with a three-legged stool for a guy to sit on while he was being "serviced." Forget about atmosphere. Five bucks for a hand job, ten for a blow.

One of the most successful, and notorious, of these clubs was the Keen Club A-Go-Go. The Keen had been around since the thirties. Every now and then it got a little face-lift—maybe some new neon to keep up with the times. But no matter what they did, it was always booze, broads, and a blow.

The Keen was run by Al Lorenz. It was a sleazy joint with loud music and girls dressed in go-go boots and sequined bikinis dancing in little cages or wiggling half-naked around the stage. No matter what time of day it was, the place was always pitch-black inside and smelled like a cross between old cigarette butts, stale beer, and piss. Very gross. And always packed.

Doc liked to joke that all of Al's girls made money "hand over fist." And it was true, too. Lorenz might have been a slimeball degenerate, but he had a real talent for making the green. He was like the P. T. Barnum of vice rackets, always dreaming up some new weird deal to attract a crowd. Some of them worked really well, and some of them were total disasters—one of his biggest being when he came up with this plan for having a live animal on stage with the dancers. He sold Doc on the deal and then went out and bought a bear. The first time I saw that bear on stage, I nearly shit a brick. I thought Doc and Al had really done it this time. I marched right over to the White Star to see Doc and told him straight out that I was worried about the dancers—and the customers.

Doc sat there, shit-faced, looking at me with those yellow eyes like he was half awake. When I was done, he poured himself a drink and said, between gulps, "Everybody's got pussy, laddie. Al says we need somethin' more to keep folks comin' in. Somethin' that'll give the Keen an edge over the competition. Now you tell me . . . who else in Cook County can say they got a fuckin' bear?"

Doc had me there. I couldn't think of anybody crazy enough to put a wild animal next to a bunch of half-naked girls—unless it was Al Lorenz and Doc Rust. It was useless trying to talk to either one of them.

From day one, the Keen had people elbow to elbow thanks to that poor bear. And just like I figured, somebody was always getting hurt. Even though the bear was declawed and Lorenz kept it muzzled and chained, there was always some asshole standing on the sidelines trying to get it riled up. Naturally, the bear would get pissed and let somebody have it—usually one of the girls on stage. But that didn't stop Doc and Al; when they got a look at the bar receipts, they figured they were onto something. Next thing you know, the Keen had a lion up there on stage with that bear. I was always getting emergency calls for assistance and an ambulance, but the joint was making money, and that was the bottom line.

Really, in the Outfit you can't let anything or anybody get in the way of making money. It's all about earning. And doing what you're told. No matter what it is. Sure, you might hate beating the hell out of some poor guy who's late on his vig (his interest payment). You might even feel sorry for the dumb bastard. But if that's what they tell you to do, you do it. No questions.

⊕ ⊕ ⊕

Of all the rackets in Willow Springs, there was nothing that could beat those machines. They were a gold mine. Aside from the slot machines—and slots were always a big moneymaker—there was one type of machine in particular that was very profitable: the bingo machine.

Bingo machines were manufactured by a Chicago company, located on Belmont, called Lion Manufacturing. Lion would later morph into Bally. Bally was selling machines to companies connected to the Outfit.

Back then, I had no idea the impact those machines—as well as Bally and the gaming industry—were going to have on my life. Who would've thought that some back alley machine company would actually have an impact on world politics. It sounds nuts, but it's true. And I was going to be right there when it happened.

Back in the sixties, the only guys who realized that Lion Manufacturing offered that type of opportunity were the Jews out of Las Vegas, Chicago's Sam Giancana and his sidekick Hy Larner, and New York's Meyer Lansky—and a handful of very ambitious world leaders and counterintelligence agents. All the average connected guys like me knew at that time was that Lion's machines had to be some of the biggest gravy trains ever invented. And I didn't care where those trains were headed. I just wanted to climb on board.

But it was going to be several years before that happened. In the meantime, I kept close tabs on the machine business. I'd been fascinated by that particular racket ever since I'd pulled my first fistful of coins out of a fifty-gallon drum at A&W. And given that connection, it was only natural that most of my early information about Lion came from A&W's Pete Altieri.

In the scheme of things, Pete was just a low-level guy, but over the years he'd made millions off those machines and poured it all into legitimate investments like theaters, restaurants, and construction projects. If there was anything Pete was an expert in, it was the machine racket, and the way he told it, William O'Donnell ran Lion Manufacturing. O'Donnell wasn't just a front for the Outfit or some sap figurehead who rigged machines. O'Donnell was very influential and, Pete said, extremely tight with some very powerful Jews, including Sam's gambling wizard, Hy Larner, and the Teamsters Union pension fund managers Red Dorfman and his son Allen. In particular, Pete said, O'Donnell was close to the well-known Las Vegas Jewish entrepreneur Hank Greenspun and his far less public Zionist friend, the arms smuggler Al Schwimmer. According to Pete, Greenspun had introduced O'Donnell to several international figures and political leaders, among them wealthy Saudi casino patron and arms dealer Adnan Khashoggi.

None of that impressed me at the time. To my limited way of thinking, Zionists and arms smuggling didn't have anything to do with international leaders and bingo machines. But what did grab my attention was what Pete said about O'Donnell's connections to the Las Vegas casino scene. Supposedly, O'Donnell even had a say in who got what job out there. If that was true, William O'Donnell wasn't just a thief, he was a very powerful thief. And a very creative one, too.

When it came to creativity, O'Donnell's bingo machines were the best example of highway robbery ever invented. He had them all set up as a no-win situation. People dropped money in them like crazy because if you hit five in a row you'd win around two hundred dollars. But did anybody ever win? Never—well, maybe once in a million years. When it did happen, it was really something. Bells went off, lights flashed. The tavern owner would give you the works. They actually threw a party in the joint. They wanted everybody to know that somebody finally won.

Pete's company, A&W, set up hundreds of O'Donnell's bingo

machines for Turk and Joe in Willow Springs, all of them rigged "no-win" and bringing in as much as a hundred grand apiece or more. With that much action, there was always somebody calling the cops and complaining about getting taken. We didn't do a damned thing to help those poor suckers, but for making it look good, you got a few bucks on the side from Turk and Joe. Pretty soon I was taking care of all those complaints. I'd go see the wife of the dumb sap who came home without his paycheck on Friday night because he dumped it all in one of O'Donnell's bingo machines. Or the guy who got his ass kicked because he lifted up the machine and was trying to manipulate the ball into the hole.

It was a real pain in the ass—but it was one more way for me to make a little extra green on the side. I never got an enormous hit—it was steal a little here, a little there, and eventually it would add up. To make a decent living in law enforcement in Cook County, you had to play all the angles. In all my years on the police force, I don't think I knew a single guy who got by on a cop's salary. The little bit of money you get paid for risking your life—it's an insult. It's like asking a guy to steal. And with a little practice—and a badge—you get real good at it, too. But you're the house, so you have it rigged. Just like those bingo machines, it's a no-win situation for John Q. Public. Cops are always going to get paid. One way or another.

⊕ ⊕ ⊕

From the very beginning, I was pretty much on my own in Willow Springs. After just four nights with Art Doogan, Kresser put me out on the street by myself, working from midnight until eight in the morning. I had a beat, with different sectors I was supposed to cover—commercial, industrial, new residential under construction, and the regular streets. On patrol you were expected to put on a hundred miles a night. I came up with plenty of innovative ways to do that without actually doing it. Usually I just pinched as many speeders as I needed to make my nut for the night, and then if I still had a few hours left, I'd go over to a gas station and have them put the car up on the rack. We'd put it in gear and let it run off a hundred or so miles and then take it down and off I'd go—to some nice quiet spot where I could catch a few z's before my shift was over. I didn't feel at all bad about doing that. It wasn't like I was going to miss anything. Most nights, Willow Springs was very quiet. It was also very boring. I had to get real creative just to stay awake out there. Sometimes I even took guys out with me in my squad car to help break up the monotony.

I took Pete Altieri on patrol a bunch of times. Usually we'd get some sandwiches, have a few laughs. He'd fill me in on the machine rackets. Eventually I even got my dad out. I had to convince him it was okay, that I wouldn't get fired for having a civilian with me while I was on duty, but once he got past that, Dad thought it was real exciting. He loved being out on the street at night with a police officer who, it just so happened, was also his son.

When Turk got wind of me taking guys out in my car, he begged me for a ride. Turk loved the whole idea of me being on the police force. And it wasn't just the car, either. Turk loved everything about it: the gun, the uniform, the badge. Turk thought I was the neatest cop he ever met in his entire life. Pretty soon it became a regular deal, me and Turk in that patrol car. We'd pretend he was my partner, and all night long we'd pull people over. We'd make traffic stops, and people would just beg him to let them go. And just when he had them in the palm of his hand, Turk would go soft. He'd get back in the car and say, "We can't give this woman a ticket. She says she's got three kids, she works four jobs, blah, blah, blah. . . ." He always fell for a sob story.

The idea of me being a cop wasn't such a big plus with Joe Ferriola. He didn't want to get within a mile of a squad car or a police station. Consequently, Joe and I didn't play around like Turk and I did. I'd see Joe for breakfast, we'd trade a few jokes. We were still friends, but it was different than before at the gas station.

Naturally, my relationship with Turk and Joe still worked both ways, just like it had at my Sunoco. I was their inside track on the police force. If there was something they needed, they knew where to go. And it was the same with all the other guys, too. Marshall Caifano dropped by now and then. Sometimes I'd run into him and

Butch with one of the other guys at a local diner. And of course, there were Vinnie Inserro and Mugsy Tortorella.

At that time, Vinnie and Mugsy were doing some big hijacks. With me on the force, they started bringing their trucks through my territory. Vinnie was constantly dropping off swag at the station. You never knew what was going to come through the door. It could be anything from fishing rods to fur coats. It almost felt like old times. But of course, it wasn't. I was a cop and this was a police station. It was mostly "Take a look at this guy's ticket for me." But I didn't mind. I figured fixing a ticket was the least I could do for a friend. Obviously, I wasn't forgetting who my friends were.

CHAPTER 14

It was sometime late in the spring of 1965—I was just getting used to being a cop—when I started hearing about a grand jury investigation that was being put together by U.S. Attorney Ed Hanrahan. According to the papers, Hanrahan and his team of federal prosecutors were planning to go after all the rackets in Cook County. It was going to be an all-out war on organized crime.

I started to worry that Willow Springs wasn't the best choice of locations for somebody in my line of work. As I said, Doc and Kresser ran their rackets wide open. I figured we were asking for trouble doing business like that with the feds snooping around, so I made it my business to talk to Doc and Kresser about what precautions we should take. But it was a big waste of time, talking to them. The papers could say all they wanted about the feds putting the squeeze on the Outfit's rackets in Cook County, but they might as well have been talking about oranges for all Doc and his pals cared. The investigation was a joke to them. And it *was* a joke, too, at least initially.

But that wasn't the case anywhere else in the county. Outside Willow Springs things were very tense. There were all kinds of rumors during the first week of the investigation. The papers were talking about some major heat coming down. All the guys were getting subpoenas. And so were their relatives, their barbers, you name it—everybody who was anybody was being called to testify before that grand jury. Everybody, that is, except Doc and his Willow Springs crew.

After a couple of weeks of nothing of any real consequence coming out of the investigation, everybody outside Willow Springs started to lighten up. It looked like the investigation was a total bust. But if anybody had stopped to think about it, the outcome of the investigation would have been predictable from the start. Hanrahan and his buddies had to know there wasn't an Outfit guy on earth who was going to spill his guts. The FBI agents involved certainly knew it was going to be the Fifth Amendment all the way. During the entire grand jury investigation, there weren't any big revelations about the underworld. Nobody learned anything new about the so-called inner workings of the Outfit and its rackets. Which meant that Hanrahan's little investigation was nothing but a bunch of bull. Or was it?

I was starting to hear through the grapevine that the grand jury was a smokescreen, that Hanrahan was really after Sam Giancana. If that was true, I figured I was headed for a few problems of my own. After all, Sam Giancana had gotten me my job as a cop. He was my chinaman. In the Outfit, you line up with a power guy and if you're lucky you get pulled along with him, wherever he goes. But if he gets whacked or pushed aside or pinched, nine times out of ten you're screwed. If the feds took Sam Giancana down—for whatever reason—it was a pretty good bet there would be a shakeup in the entire organization.

But there was another thing that had me even more concerned about my ties to Sam: for some time he'd been getting a lot of pressure from *inside* the organization. Granted, a boss will always have to take some heat, but this was more than that. Word had it that the boys at the top were having some serious doubts about Sam's ability to handle things. And it wasn't just the top-echelon guys who were on his case, either. He was getting it from all directions.

There were the young turks like Tony Spilotro who thought they were being held back and were starting to make some noise. And the middle management types who wanted a cut of Sam's overseas gambling operations. And then there were the old-timers like Tony Accardo who carried a lot of weight and wanted Sam to knock off all his high-profile celebrity jet-setting because they were convinced it was hurting business.

At least some of those issues came to a head during the fall of 1964, when the top men in the Outfit held a major sit-down at the Armory Lounge. When they walked out, it had been decided that Sam's old 42 Gang pal "Teets" Battaglia was going to be handling the Outfit's day-to-day operations from now on. Some people thought that meant Sam had been pushed out and stripped of his power, which Turk and Joe said was exactly what Accardo and his pals wanted people to think. If Teets Battaglia was considered the boss, Turk said, then the heat would be off Sam, and everybody could get back to business.

It probably looked like a good plan at the time, but six months later, in May of 1965, when the feds called "the boss of Chicago's underworld" to testify before the grand jury, it was pretty obvious they hadn't fooled anybody. Instead of Battaglia, Hanrahan and his colleagues called Sam Giancana to take the stand.

Naturally, Sam took the Fifth. But the prosecutors didn't act at all discouraged. They kept firing questions, dozens and dozens of them. And it soon became obvious that Hanrahan had been after Sam all along. He'd spent thousands of taxpayer dollars on the effort, too—a fact substantiated by Bill Roemer in his autobiography, *Roemer: Man Against the Mob.* In his book Roemer talks about the "novel strategy" used in 1965 to bring down Sam Giancana, calling it "the brainchild" of the Justice Department. According to Roemer, Hanrahan had been laying his trap with all those questions. He wanted Sam in a real tight box when the feds made their play. And what a play it was. They led Sam out of the jury room and took him and his attorney before U.S. District Judge William Campbell. It was time for the grand slam: Judge Campbell offered the boss of Chicago's Outfit full immunity.

By law, Sam had to accept the offer of immunity or risk being found in contempt. No mob boss in history had ever been offered full immunity. In order to do that, the government prosecutors had to be willing to overlook Sam's past crimes. *All* of them. With contempt charges hanging over him, Sam didn't have much choice. He had to accept their offer, but there was no way he was going to talk. He got back on the stand and did what anyone else in his position would do: after giving his name and address, he clammed up and took the Fifth. That meant he was in contempt, which was what Hanrahan was waiting for. Within minutes, Sam was handcuffed and hauled off to the Cook County Jail—where the judge said he'd stay until he agreed to talk. When word got out that Chicago's boss was behind bars the whole town went crazy. Nobody could believe they'd put Sam Giancana in the slammer.

The grand jury continued after Sam was jailed, but every day it was looking more and more like the investigation was dead. Then, out of nowhere, the *Sun-Times* and *Tribune* started to hint that the feds were going to grant immunity to all the top Outfit guys and put the entire Outfit behind bars.

One night I dropped by the White Star with a copy of the paper. Doc took one look and threw it in the trash.

"Bullshit," he said. "It's gettin' to be where the newspaper's not fit to wipe a man's ass. All that crap about grantin' the rest of the Outfit boys immunity . . . that's just one more sorry-ass reporter tryin' to sell a fuckin' paper and make a name for himself. Laddie, they wanted Sam. And now they got him . . . by the fuckin' gonads, too, from the looks of things."

The way I saw it, they didn't just have Sam Giancana by the balls—those feds had a pretty good grip on me, too.

⊕ ⊕ ⊕

As a rookie cop in 1966, I could never have imagined what Sam Giancana and his friends in the CIA were up to. Then again, it was next to impossible for someone in my spot to have that type of information. The nitty-gritty details were pretty much out of reach. Generally speaking, I only saw parts of the whole or maybe a few broad strokes.

From what I heard on the street that winter of 1966, even

though Sam was still in the Cook County Jail, he'd managed to put together some major deals. Of course, I had no idea how major they were. Or what they'd mean down the road. I picked up most of my information from one of the guys—mainly Pete or Turk and Joe. Sometimes I'd get wind of something through Butch Blasi when he dropped by the village. Occasionally I might even learn a thing or two from Mugsy or Vinnie. But wherever my information came from, it was never complete.

Typically, in the Outfit, only one or two individuals ever have the full story on a situation. Doing business like that keeps some snitch from knowing enough to bring down the entire operation. So most guys just know one particular aspect of a deal. They have no idea how it all fits together. For example, Pete Altieri knew everything about running a machine operation in Cook County, but his knowledge was limited about the machines outside his territory. There were a few people—like Doc Rust and Mugsy Tortorella—who were a bit crazy but somehow always had an ear to the ground. They always knew a little something about everything, not because they were in the top echelon but because they weren't tied to one line of work.

Hustlers, like Mugsy, were into anything that made a dollar. No matter where it was, if they smelled an opportunity, they'd be there. Mugsy usually had the inside scoop, or at least part of it, on just about anything you could think of. Like Doc, he knew people all over the place, so he always had a line on what was going down somewhere. And like Doc, Mugsy was always good for a story. Which was how I learned, late one winter night in 1966, that Sam— who was still behind bars at the time—was involved in a dope-smuggling racket.

We'd been riding around in my squad car, killing time, when Mugsy started talking about how much money there was in dope. I had no idea any of the guys were into drugs; to say I was surprised would be a huge understatement. But I didn't want to look like an idiot, so I said—in this casual, offhand sort of way—that I'd always heard dope was strictly off-limits.

"Are you kiddin'?" Mugsy said, like he thought I was nuts.

"Hey, what the fuck do I know. I'm just a goddamned cop, right?" I shrugged. "But the word I've always heard is that messing around with dope can get you whacked."

Mugsy shook his head. "Not if you keep it quiet and don't flaunt it around the old-timers." He pulled a thin, rolled-up cigarette out of his coat pocket. Marijuana. At the look on my face, Mugsy rolled his eyes. "Jesus, kid, relax, will you? You need to get out of this fuckin' squad car more often. I mean, where the hell you been? All this cop business is turning you into a real square." He lit it and took a deep drag. "Man, wake up," he said, exhaling a cloud of smoke. "I mean, this is the fuckin' sixties."

I could feel my blood pressure rising. Ever since I'd gotten a badge, I'd taken a lot of crap from people who knew me from when I had the gas station. They were always on me about being a cop. It was supposed to be good-natured ribbing, but after a while it got under my skin. It was like being a police officer made you an automatic square. I knew what goddamned decade it was.

Mugsy switched the radio from Herb Alpert's "A Taste of Honey" to the Rolling Stones' "I Can't Get No Satisfaction" and handed me the joint. I'd never smoked reefer before. I was a cop and marijuana was illegal, so as far as I was concerned that made it off-limits—not square—which, I admit, doesn't make any sense given my other activities. But there were some lines I hadn't planned on crossing, and smoking dope was one of them. I considered marijuana like booze in that it could put you out of the deal. And the last thing I ever wanted to be was out of control. At that time I hadn't figured out that life's not about a plan. But plan or not, there I was, stuck. I was in a bad spot. I took that joint, put it to my lips, and inhaled.

That night I crossed the line. But I also got my first real look at what Sam Giancana and the Chicago Outfit had going behind the scenes. The way Mugsy told it, the Outfit was into more than pull tabs and pinball machines; they'd been smuggling dope into the states since the late fifties. According to Mugsy, Sam—along with Meyer Lansky, Santo Trafficante, and Carlos Marcello—had set up a seafood import company in Miami as a front for their drug-trafficking operation. Initially they used Marcello's contacts in Guatemala to get things off the ground. Then they started moving into the heroin trade (something I later confirmed)—which was a natural, since each one of them had excellent connections for opium in different parts of the world. Mugsy said it wasn't long after that that the mob was moving in on the traditional heroin traffickers; they were even taking over the poppy fields in some places. "Someday," Mugsy said, "they'll own it all."

By the time Castro took Havana in 1959, Mugsy said, things were going so well that Sam had pulled Johnny Roselli out of Las Vegas and put him in charge of the Miami seafood operation. Roselli was an old Chicago standby who'd worked Las Vegas and Hollywood for years. He knew how to make the big bucks. Mugsy said Roselli also knew all the right people, from politicians and corporate executives to intelligence guys and top brass in the U.S. military—a real trick for somebody in organized crime.

Johnny Roselli was very smooth, a real Hollywood type. He was also known as an excellent deal-maker and negotiator—two qualities, Mugsy said, that came in real handy when the CIA turned to organized crime for help after Congress refused to fund their covert plan to overthrow Castro.

It wasn't any secret that the mob wanted back in Havana every bit as much as the U.S. government. Organized crime had lost a fortune thanks to Castro shutting down their casinos and rackets. So I didn't think Roselli had to work too hard to hammer out a deal between the CIA and the Outfit. All he probably had to do was deliver the message to Sam. For guys like Sam and Meyer Lansky, it must've looked like Johnny Roselli had laid the opportunity of a lifetime at their feet.

From that point on, Mugsy said, organized crime and U.S. intelligence started working together, using the Outfit's Miami import company as home base for a drug- and arms-trafficking operation. No matter where they were in the world and what they were up to, the mob's rackets had the full protection of the U.S. military and the CIA.

But the Outfit didn't get that type of treatment just because somebody in the CIA liked them. Mugsy said it came with a price: Sam Giancana and his pals had to share the profits with some rogue guys inside the CIA. Still, not a bad deal considering how much more they were probably making without the law on their backs. And Mugsy said there was another thing that had made the Miami front even more profitable since they'd started working with the government: they weren't just peddling dope anymore. They'd expanded. They were running guns now, too. Guns that the CIA was buying with its share of the dope profits.

As I later learned, the idea in the beginning was that the Outfit and the CIA would work together temporarily, just until they'd gotten rid of Castro. The Outfit would help arm the mercenaries and Cuban exiles who the CIA was secretly training to invade Cuba, and that would be the end of it. Mission accomplished. But when the CIA's military invasion at the Bay of Pigs went south, that wasn't the end of it.

According to Mugsy, those CIA boys couldn't resist getting in deeper. I imagine they told themselves there'd always be another Castro out there in some other country who they'd want to knock off and couldn't because of Congress and American law. But with the mob as their partner, they had to figure it would be a whole different ball game. They could eliminate anybody who got in their way. They'd be free to target world leaders for contract hits—something they'd actually done, *unsuccessfully*, in the case of Fidel Castro and, from what I've been told by other men inside the Outfit, *successfully*, in the case of Jack Kennedy.

There was also the fact that there would always be some third-world government the CIA would want to prop up or tear down but couldn't—because the American people wouldn't approve of them meddling in other countries' affairs. But with all the dough that came from working with organized crime, they wouldn't need the public's approval with its taxpayer money to fund their little war games. They wouldn't need anybody but Sam Giancana and his connections all over the world for murder, dope, and illegal arms.

According to Mugsy, it wasn't Sam Giancana who'd put the deal together with the CIA. And it wasn't Roselli. Roselli was a good front man, Mugsy said, but he wasn't "a fuckin' genius when it came to makin' the green." Mugsy told me it was "Sam's man in Central and South America"—Hy Larner—who was behind it all. Larner had been raking in the dough for Chicago's gambling and smuggling operations all over the world, and it was his idea, Mugsy said, to move the Miami operation to Central America, where in the late fifties he'd stumbled on a friendly little country called Panama.

Thanks to the Outfit's gambling ventures in Panama, Larner had uncovered a number of friendly banks and corruptible leaders eager to do business with rich Americans. Under his supervision, every dime the Outfit skimmed off their gambling rackets all over the world had been flown into Panama. Each week, millions in cash were either deposited in the local banks or flown back to the United States via Las Vegas. Evidently Hy Larner realized that the CIA's biggest problem wasn't going to be keeping its dealings with organized crime under wraps. Its biggest challenge would be laundering all that dirty money. And it was drowning in cash until, like Hy Larner suggested, it moved the smuggling operation from Miami to Panama, where the banks, for a price, were more than willing to cooperate with criminals and American officials.

After they got settled in Panama, the partners continued smuggling dope and weapons. In fact, Mugsy said, things had gone gangbusters thanks to President Johnson's escalation of U.S. military involvement in Vietnam. All Trafficante's boys had to do then was point out an opium lord and the U.S. military would march in, eliminate the poor bastard, and take over his poppy fields. In just a few short years, the CIA had managed to stuff thousands of body bags with millions of dollars in heroin and, under cover of the military, airlift them out of Vietnam and into the hands of organized crime. Talk about a sweet deal.

By the time 1966 had rolled around, the CIA and organized crime had gone way past your typical smuggling operation. As I'd later discover, things had turned a lot more political—and a lot more dangerous. At the insistence of Meyer Lansky, Sam and his pals started working with the Israeli Mossad, smuggling weapons into the Middle East. Everything was coming in and out of Panama, which meant that everything was being handled by Hy Larner. Larner was without a doubt Sam Giancana's most trusted financial adviser. He had everybody who was anybody in Panama—from bankers to generals—eating out of his hand. Once they started running guns to Israel, Larner also had the U.S. military and its airstrips at his disposal. It wasn't hard to understand why Hy Larner was Sam Giancana's golden boy.

Over time, Larner got tight with other bosses. He probably had more business dealings on a day-to-day basis with Trafficante and Marcello than he did with Sam, especially since he'd been spending most of his time down south in Miami or Panama City. While down there he'd become very friendly with a number of world leaders and key players in the CIA and U.S. military. He was also tight with the guys in Las Vegas—money types like the Teamsters' Allen Dorfman and the Vegas media mogul Hank Greenspun.

But it was Meyer Lansky who Hy Larner got closest to. He and Lansky had a lot in common. They were both absolutely brilliant when it came to handling money, probably the best the mob had ever seen. They were also Zionists, passionate defenders of the divine right of Jews to occupy the Holy Land of Jerusalem. You wouldn't think a person's religion would make any difference, not when it came to a deal like the one organized crime had with the CIA. But Hy Larner and Meyer Lansky weren't just Zionists, they were also mobsters who believed the end justifies the means. Put organized crime and the U.S. government at their disposal and you've got a very powerful force, a force that's capable of changing the face of the mob. And the world.

⊕ ⊕ ⊕

Whenever you have a conversation with guys like Mugsy Tortorella, the discussion always gets around to money. That's a hustler's thing, so it was only natural that Mugsy would eventually bring up Sam Giancana's finances. And according to him, even though Sam was behind bars, he was making a fortune. To prove his point, Mugsy said he'd heard that the skim from Sam's Iranian casino alone was bringing in over three hundred thousand dollars a day. "That's nine million greenbacks a month," Mugsy said. "From just one fuckin' deal. And he's got shit goin' all over the fuckin' world. Man oh man, Sam's rollin' in it."

Mugsy said there was so much money coming through Panama now that it was even starting to make its way down the line to guys like Sam's driver and bodyguard, Butch Blasi. Word had it that just for playing Sam's delivery boy on the Panama deal, Blasi alone was taking home twenty-five thousand a month.

I was speechless. I couldn't believe somebody like Butch Blasi was bringing down that kind of dough. So he carried a few bags of money for his boss now and then. Big deal. And he was a nice guy and all, but he wasn't exactly a goddamned genius. Butch Blasi was a bodyguard, a driver, for Christ's sake.

Now I wasn't a hustler like Mugsy Tortorella, but I imagine our thoughts were running along the same line that night. Yeah, we both had just one question: Where's my bag of money?

⊕ ⊕ ⊕

Hearing about Blasi making all that dough really brought me down. Looking back now, I see that it was another turning point for me. I had to admit that my career—my entire life—was at a standstill. I wondered if I'd ever make it with those guys. I got very low. Every morning I'd get up and look in the mirror and ask myself, Where's my future? Where's the bigger payday? I reminded myself that I was bringing down sixty grand a year on the side as a cop, and that was a lot of dough. I also had another forty or fifty from racing. But a hundred grand a year was a long way from Blasi's twenty-five grand a month. And Willow Springs, Illinois, was a long way from Panama and all those big deals.

I was also pretty sure I wasn't headed for any big promotion on the force. No way was Doc, cheap bastard that he was, going to pay me more. What did it matter that I was doing a bang-up job scamming all those saps who exceeded the speed limit? And so I got a little extra out of Turk and Joe each week, so what? Most of my dough—my bread and butter—was coming from Mugsy's hijacks. I had to face facts: my opportunities to make a buck in Willow Springs were limited, and they could get more so at any time.

Deep down I knew it was time to move on. But before I could do that, I had to follow one of those unspoken rules of Outfit etiquette: I needed Sam Giancana's permission. And he was sitting in jail.

CHAPTER 15

I'd been hearing that Hanrahan was planning to put Sam back on the stand before the grand jury was terminated in May. Supposedly he was again going to offer immunity. If Sam continued to take the Fifth, he'd stay in jail—for another eighteen months. If he cooperated, he'd walk. But walk where? The odds of him being alive twenty-four hours after that wouldn't make for a decent game of Russian roulette. Sam Giancana would be a dead man. Everybody knew that, including him. With those options, I figured Sam was going to be behind bars for long time.

My entire future was locked up in the Cook County Jail and I didn't have the slightest idea how long it would stay that way. I didn't know what to do. All I knew for sure was that sixty grand a year wasn't going to be enough, and that without Sam's blessing I couldn't leave the force in Willow Springs. I decided to bite the bullet and got a second job, driving a truck for a gasoline company on the nights I was off patrol. When I realized that still wouldn't cut it, I got a third job, working days at a gun shop where I'd been doing target practice.

Even with three jobs and all that dough on the side, it didn't matter—there still wasn't enough to go around. And it wasn't like I was in debt, either. I handled my finances very well. I lived it up now and then, but I wasn't stupid. And I wasn't selfish. I took care of my folks, my friends, sometimes I even took care of a total stranger. But to do all that, I was working around the clock. I had a million balls in the air. Things were real crazy, not only in my work but also in my personal life.

When I wasn't sleeping and wasn't working, I was racing cars. I was never home. I was more like a stranger than a husband and father. I had to work all those jobs to make ends meet, and I could justify that, no problem. But when it came to my free time, I knew I had a choice. And the truth was, I wasn't about to give up racing. Besides it being the love of my life, it represented a substantial amount of income. I'd locked in a top-notch sponsor, Grand Spaulding Dodge. My picture was in all the papers. Locally, it was like Michael Corbitt was this famous race car driver. For a guy who was at a low point in most other parts of his life, that sort of attention felt pretty good.

Maybe it was all those race car groupies hanging all over me every time I stood in the winner's circle. Or maybe my career situation being what it was made me hungry for a little excitement. But for the first time since I got married in 1961, five years before, I started playing around on the side. I don't know if Annette knew what was going on, but she was very unhappy. We'd fight and then we'd make up. She'd move out and then move back in. We separated, we went back together. Back and forth, over and over. It was driving us both crazy. We were having some very tough times.

Annette felt very strongly that my way of life was bringing us down. She was always giving me the old money-isn't-everything song and dance. She didn't want me to spend all my time working three jobs, and she definitely didn't like me being a cop, with all those nights away from home. Looking back now, I know she only wanted what most people want—somebody to love her, a decent family life. But at the time, that wasn't my thing. I wasn't the type to sit at home in front of the TV, rocking a kid in my lap. I'd never been able to sit still for five minutes, I craved action—not exactly the qualities of a family man.

⊕ ⊕ ⊕

It was sometime that spring that it finally hit me that it wasn't just money that was driving my life down a certain road. Action was almost as important to me as having some green in my pocket. And the truth was, when it came to being a cop in Willow Springs, there was nothing happening. Not to say I hadn't known that from the beginning.

Chief Kresser had actually laughed when I told him I was taking instruction in the use of my weapon. "Why?" he said, with a straight face. "This ain't the movies, kid. There's no call for any quick-draw gunplay out here in Willow Springs. We got no crime."

In any case, after a year of practicing at the target range, it was all automatic for me. I didn't have to think about how to handle a weapon anymore. I knew that if I ever needed to use my gun, I'd react properly to a given situation instead of pissing my pants. By that time I'd realized Kresser was right—there wasn't much occasion to use a weapon in Willow Springs. And from the looks of things, there probably never would be. Since I'd been on the force, the only time I'd even had to take my gun out of the holster was to clean it or put it away. Most of the calls I got were from the country-and-western shit-kicker joints reporting some parking lot fight.

There was one exception. One night in early May of 1966, a motorcycle gang went into a country-and-western tavern and, for no reason at all, started tearing the place up. There were maybe fifteen or twenty big Hells Angels wanna-bes wearing leather and chains. They were carrying baseball bats, crowbars—all sorts of weapons. By the time I got the call, the bartender said a few of the customers had been hurt, some seriously.

Because there were a lot of little towns like Willow Springs in the county—towns that had just one squad car on the street—we'd all gotten together in a mutual-aid pact. That meant if one cop called, the others came running. When I put the call out, six squad cars descended on that joint.

From the minute we walked in the door, it was a free-for-all. Those gang members were hitting us with chairs, bottles, pool cues, whatever they could lay their hands on. Then the brass knuckles and switchblades came out. It was wild. I got my shirt torn right off my back, and then my watch disappeared. Two guys even tried to take my gun away. It was total chaos. I was loving it, too. We gave as good as we got. At the time, we didn't have a rule book on how to deal with a situation like that so we just followed our instincts. In other words, we beat the ever-loving crap out of them. We ended up with quite a few severely beaten individuals on our hands. I got on the radio and called hospitals all over the area, and pretty soon there were a bunch of ambulances screaming in, sirens going and lights flashing.

When it was all over, the tavern was trashed. I started looking around for my watch, but I wasn't going to find it in that mess. I was pissed about losing it, too. But that wasn't all—my uniform shirt was ruined, which meant I was going to have to cough up the dough for a new one. Forget about the fact that I was black and blue and had big chunks of hair pulled out of my head—forget all that. It was that goddamned uniform shirt and watch that had me worked up. Later, when I told Doc what happened, he acted like it was no big deal. "Don't worry, laddie," he said. "We'll take care of you." I wasn't going to hold my breath.

Two weeks went by and I got a call to go to the Keen Club. I zipped over and saw Turk's big black Cadillac parked by a phone booth. I hadn't seen Turk and Joe since before the fight at the tavern with the motorcycle gang, and I was still looking pretty rough. I got out of my squad car and those two guys took one look at the black eye I was still sporting and started to laugh. Joe said, "What the hell happened to you?"

"What happened to me?" I said. "What the fuck do you think happened? That friggin' joint of yours down the street, that's what." I told them the place was nothing but a menace to society, that somebody should close it down. Of course, reacting like that only made them laugh that much harder.

They must've figured they'd better knock it off before things got out of hand, because all of sudden Joe stopped laughing and got real serious. He said, "Relax, man. We're just fuckin' with you." He said that Doc had told them all about the fight at the tavern and how I'd taken a pretty good beating. He looked over at Turk and winked, and Turk handed me a paper bag. "Here's a little somethin' to let you know how much we appreciate what you're doin' out here," Turk said.

I felt like a real jerk, losing my temper like that. I managed to thank them before they left, but I knew that wasn't enough. And I felt even worse when I opened the bag and found a thousand dollars and an Omega watch. Turk and Joe always treated me right. Really, they were like brothers to me. Yeah, I guess you could say I loved the guys.

⊕ ⊕ ⊕

I wasn't talking. And I didn't feel like eating, either. I just sat there and watched Vinnie Inserro rub a limp french fry back and forth across a glob of catsup on his plate. "Yeah," he said, "that's right. . . . Sam wants Turk and Joe out of Willow Springs."

I looked around the diner for a second and down at the hamburger in front of me. The joint had the best burgers around, but I'd lost my appetite. "So they're moving?" I said, trying to smile.

"Yeah, they're movin' all right. Sam wants Turk workin' Cicero with Joey Aiuppa. He's sendin' Joe over to handle things on the North and West Side. They're movin' up, the lucky bastards."

Given my feelings about Turk and Joe, you can imagine my reaction to that piece of news. When Vinnie had suggested we meet at the Four Es for lunch, I'd figured he was going to fill me in on some new deal Turk and Joe were setting up, not introduce me to their replacements. But that's exactly what happened. After Vinnie dropped the little bomb about Turk and Joe, two men walked in and headed straight for our table.

As they sat down, Vinnie said, "Mike, these are the guys that are takin' over out here for Turk and Joe . . . Tony Orlando and Paul

Kelly." We shook hands and I tried to be friendly, but they were very odd. They were older than most of the Outfit guys—maybe in their late sixties—and to me they looked like midgets. Orlando was maybe five-two. The other guy, whose name was actually Payne, not Kelly, was probably even shorter than that. Overall they were both very peculiar. Physically, they reminded me of trolls.

Other than being so old and so small, there wasn't much about Orlando that was particularly memorable. But one look at Payne and you'd never forget him. Paul Payne was a Jew who had been an operator for Hy Larner for years. He even had the same nickname as Larner—"Red." And talk about weird. Payne had real red lips, like he had on lipstick. And he had this very pale, very white skin—which made his mouth look even redder. He was also almost totally deaf and wore a big hearing aid in each ear. It's hard not to gawk at somebody who looks like that, especially when he's sitting across the table.

Fortunately, Payne did all the talking. It was clear he was the top dog in the operation. He handed me a business card that read "Multimedia Advertising and PR. One hundred enterprises of all kinds" and said they were taking over Willow Springs. Up until that day at the Four Es, I'd never even heard of Paul Kelly or Tony Orlando, so I wasn't at all impressed by the fact that they had a business card. But later, when I found out how big their operation was, I gained a lot more respect for them. Believe it or not, those two old guys controlled all the gambling devices, pinball machines, and so on in Justice, Hodgkins, Hickory Hills, and Lamont. And with Turk and Joe out of the picture, they were adding Willow Springs to that list.

They might look like freak nobodies, but they were richer than King Solomon—and very powerful. Their territory should have fallen under Joey Aiuppa, out of Melrose Park, but instead they reported directly to Hy Larner. Doc had his own explanation for that. He said, "Payne's a Jew. Larner's a Jew. No fuckin' mystery there, laddie. Them Jews all stick together."

⊕ ⊕ ⊕

Like I've said, Doc Rust and Chief Kresser made a hell of a crime-fighting team. They played all the angles. Fuck their mother, their sister, fuck you, me—it didn't matter to them. It was anything for a dime. Which was why I was instantly suspicious when Kresser called me one morning and said, "Listen, somethin's come up. I'm gonna be out of the office for a while. Doc wants you to start comin' in during the day. We don't expect you to work nights too. So I'll put on a part-timer to handle your patrol."

Kresser didn't say what he was going to be doing, and I didn't ask. I came in the next day, but I hated working days. I was a night-shift-type guy. Still, what could I say? Doc wanted me in during the day and that was that. The minute I hit the station, Kresser took an unmarked car and left me in charge, like I was the chief. He was acting way too mysterious, so as I said, I suspected he and Doc were up to something.

This went on for several days, until I decided to tail the old guy. Kresser had never been any good at spotting a tail, so he didn't notice me driving behind him all the way up Mannheim Road. He didn't even see me when he turned off and headed down a deserted old roadway that led into the Illinois and Michigan Canal. I followed him until he stopped by the canal and then I parked behind some thickets. I sat there all day, watching Kresser watch these big dump trucks come and go. Finally I realized he was counting the trucks.

As it turned out, my instincts had been on target. Doc had gotten together with the mayor from the town of Justice, and they'd given this one trucking company permission to dump in the I&M Canal. Kresser was supposed to keep tabs on how many trucks came through so they could figure how much the trucking company owed them. Doc and Kresser camped out there for months, and eventually Kresser let me in on the deal. I guess I was as shameless as they were.

I was at the canal with him and Doc one afternoon when a big black Olds drove up. It was payday. And talk about the dough. That guy got out of the car with two big stacks of hundreds. You should have seen the smile on Doc's face when he saw that money. He took all that dough like it was nothing. I'd never seen Doc—or Kresser—so happy.

The I&M Canal was supposed to serve as a runoff area for Willow Springs and was originally built to protect the community from flooding, but I didn't know that at the time. Before long we'd turned that canal into a landfill. Screw everybody else. After almost a year of them dumping in the canal, I was standing in the bathroom shaving one morning when I heard on the radio that there was a major fire in Willow Springs. *In the I&M Canal.* And it was burning out of control. All of a sudden I started wondering; there was no telling what chemicals they'd dumped in there, along with God knows what else.

As if that wasn't bad enough—having this big toxic fire burning out of control—it started to rain. Evidently Kresser had had the area backfilled with dirt so nobody could tell what we'd been up to. But all that dirt clogged up the canal—which meant there wasn't any drainage. There was no more runoff area. The entire village of Willow Springs began to flood.

Naturally, there was a major investigation. Everybody was out there at that canal: the EPA, the FBI, you name it. But Doc didn't sweat it. He'd made several hundred thousand dollars in dumping fees, so he figured he'd just pay everybody off.

For whatever reason, nothing ever came of that investigation. The Army Corps of Engineers had to spend a million dollars to put the fire out and straighten the mess up. I went out to watch the cleanup a couple of times. It was unbelievable. They were pulling all kinds of stuff out of that canal for months: steel beams, rebar, concrete, wood, carburetors, even whole cars. And Doc and Kresser? They were on to their next scam by that time. They never even looked back.

⊕ ⊕ ⊕

I admit I made some dough on the canal deal. Doc managed to throw a few bones my way. I was actually starting to feel better about being a cop in Willow Springs, and when I got a call from Vinnie Inserro, I thought things might get even better. He sounded real excited. He wanted to meet me at the Keen, right away.

I figured this was important, like maybe this was the big deal I'd been waiting for. I was on cloud nine all the way over there. I even started thinking about how I might be able to quit those other jobs and turn into a decent husband. I was in dreamland all right. Vinnie Inserro took the wind out of my sails the minute I sat down. "Mugsy's had a major change of direction," he said. "Next week he starts makin' runs between Miami, Las Vegas, and New Orleans. This is high up in the organization. We're talkin' some big money, Mike. Big money . . ."

I smiled and nodded. Butch Blasi's twenty-five grand came to mind. "So how can I help out?" I asked. My heart was doing a drumroll. I was hoping Mugsy had decided to cut me in. And then it came. The kick in the balls.

"Well, really you can't do anything . . . mostly just be patient. Yeah, that's the best thing to do. The trucking business is gonna dry up for a while. With Mugsy gone we'll be taking them through Summit. So for now, relax."

I was speechless. At that point those hijacked trucks of Mugsy's represented most of my side income. Where else could I find that type of opportunity? I'd been making a minimum of five hundred bucks a week just for escorting a goddamned truck through a two-bit town. Where would I find another deal like that?

I guess it shouldn't have been a big surprise. Hadn't the same thing happened with the gas station? And as far as Mugsy was concerned, after our recent conversation, I should've seen the writing on the wall. Mugsy had to be the Outfit's biggest opportunist. I'd seen the look in his eyes when he told me about all the dough Blasi was getting on the Panama deal.

For days after that, I kept to myself. I was crazy mad. I started scheming, and two weeks later, by the end of May, I'd made up my mind that I was getting out of Willow Springs. I decided to approach Butch Blasi, who'd been Sam's eyes and ears ever since he'd been in jail. I figured he'd talk to Sam, and maybe they'd see the light.

I had my entire speech worked out. But Butch didn't make it out to Willow Springs that week. He and Sam had more important things to deal with than some punk kid's job dissatisfaction. On

Memorial Day 1966 Sam Giancana was released from the Cook County Jail. And then he vanished.

⊕ ⊕ ⊕

The word on the street was that Sam was out of the country. For good. He was in Mexico. Or Panama. Or Iran. We were all guessing. I found out later that even the crack FBI agents tailing him lost him virtually the minute he went out the jailhouse door.

Even though Sam was out of the country, he was still the boss. So when I got an offer from the chief of police in Summit to go to work on his force over there, I felt I ought to run it past Butch Blasi. Fortunately, Butch saw the potential. He told me to take the job. He said there were things I could do for them in Summit and that I should continue to keep an eye on Willow Springs.

I went to see Doc at the tavern. I walked in, and it was like the first time I'd met him. There he was, sitting at that table with his bottle of brandy. I got a lump in my throat just seeing him like that. It was going to be hard to break the news. By now I actually think I liked the old bastard.

Doc took one look at me and knew something was up. He said, "You got somethin' on your mind, laddie? Spit it out."

"Yeah, Doc, I do. And I don't know any other way to say it than to come right out with it. I'm leavin', Doc. I'm goin' to Summit. They want me to be on the force over there." Doc started to say something, but I held up my hand. "Now hold on. Before you fly off the handle, let me finish. You of all people should understand what I'm doin' here. Hell, you're the one that always told me a man's gotta better himself when he gets the chance. And that's what I'll be doin' goin' to Summit. I mean there's more benefits over there. And the pay's better, too."

Doc just looked at me. He took a drink. His face turned all purple. He took another drink. I hadn't expected him to get so upset. "Now you listen here, laddie," he said. "You ain't goin' nowhere. You're gonna stay right here."

I shook my head.

Doc's expression changed. His voice was pleading. "But I don't want you to go, laddie," he said. "Hell, I need you out here." He gulped at his drink and then went on, "Tell you what I'll do. I'll make a deal with you right here and now. I'm runnin' for election in two years. You know I'll win. I promise you right now I'll make you chief when I'm elected. And you know I'm as good as my word." Then, like he was reading my mind, he added, "Now you may be thinking what about Joe Kresser. Hell, we both know he's through. Joe's a good old boy, but he's ready for retirement." Doc's yellow eyes came to life, and he slapped me on the back. "So there you are. . . . When I get reelected, I'll get him out and make you the new chief."

I hadn't expected Doc's reaction, and with that sort of promise, I decided I should give my decision more thought. But after a week, I still hadn't changed my mind. Even if Doc was as good as his word, the old guy could die anytime. Besides, I couldn't overlook the money and benefits at Summit—two grand more a year, a pension plan, health care benefits, and job security. In Willow Springs I had nothing. Nothing but the same old boring routine.

CHAPTER 16

For a guy who craved action, working on the force in Summit was a dream come true. In my eighteen months there I saw more crime activity than Chief Kresser saw in a lifetime in Willow Springs. We had bar shootings, firebombings, high-speed chases, you name it.

Summit was a gritty blue-collar town of eleven thousand. It was a largely white community; the blacks all lived in a crummy area on the south side. Because of the factories, the air smelled bad most of the time. Like Willow Springs at that time, the town was very corrupt, but there weren't the strip clubs in Summit that we'd had in Willow Springs. Still, half the businesses in the community were taverns, package liquor stores, and gin mills, so there was always some type of gambling going on. The big thing Summit had going for it was its police station, which was new and very modern.

Since I'd grown up in Summit, I knew everybody in town. Most of the cops on the force were the same guys who'd chased me around as a kid. When I'd had the Sunoco, we'd gotten pretty friendly. And now here I was, a fellow cop, working shoulder to shoulder with them.

I'd also known the chief of police, Sam Boscovitch, since my old

A&W days. He and Pete Altieri were very close friends, and it was that connection from the past that had led to my being hired. Boscovitch knew my rep and thought I was made to order.

Right away I got things off and running for the Outfit. I got in touch with Paul Payne and Tony Orlando and they started moving their machines into Summit. Of course, the guys in the department had no idea what I was up to. They had their own deals going, mostly truck rip-offs and security deals. Me, I was freelancing. But Sam Boscovitch was my boss, and there were some things I had to do his way, which was how I met the attorney they called "Mr. Fixit," Alan Masters. Boscovitch set up a meeting so I could get in on the Masters gravy train.

And what a gravy train it was. Everybody said that if a cop played ball with Alan Masters, he'd make a fortune in kickbacks. The way it worked was simple: every cop gave out tickets and made arrests. And every person who ever got a ticket or got arrested was always wishing there was someone who could get him off the hook. That's where Alan Masters came in. Boscovitch told me that Masters could fix *anything*—from that pesky little DWI to the murder of your wife's boyfriend. He had some judges in his pocket, some politicians. He knew who to pay off and how much it would take. All a cop had to do was refer those desperate offenders and Masters took care of everything. For the referral, the cop got a piece of the payoff. In law books they call that extortion, bribery, and racketeering, but in Chicago everybody did it. Hardly anyone ever got caught. And if they did, it was a slap on the wrist.

Unfortunately, from the minute Alan Masters walked into the restaurant for our first meeting, I despised him. He swaggered over to the table like a big shot and started throwing his weight around. He was arrogant, fat, and pushy. I was not impressed. But when he gave me his pitch and got to my favorite subject—money—I decided it might be in my best interest to overlook his personal shortcomings. The deal was just too sweet. For referring people to his office, Masters promised me a third of his fee. It could be hundreds or thousands, but it would always be in cash. Whatever bad feelings I had about Alan Masters, I ignored. Like always, it was greed that took me down the wrong road. All I did was follow.

⊕ ⊕ ⊕

I was making so much money at Summit that sometimes I forgot to cash my paycheck. I was working all kinds of deals—a lot of them with Alan Masters—and I was still taking care of my old pals in Willow Springs. I'd hear about a wire some county cops were putting in some joint over there, and I'd tell the owner to watch his step for a while. Or I'd get wind of some plan the county had to raid a card game, and I'd get to the right guys and they'd lay low for a while until the coast was clear. Then it was business as usual. Everyone in Willow Springs was very appreciative, so I got a lot of envelopes.

And as far as Summit was concerned, I'd never seen so much rooting—shakedown, extortion, bribery. But I wasn't ignoring my duties as a cop. The badge and the gun were very real for me. I was turning into Mr. Crime-Fighter Supreme. The minute Boscovitch put me on nights and gave me an unmarked car, I was up to my ass in alligators. I was a one-man gang looking for trouble. And I was very well armed. I carried a .357 magnum in my regular holster and a second gun, a .45, on my ankle strap or in the small of my back. I had a shotgun and a sawed-off semiautomatic carbine in the floorboard of the squad car. Before long, I had plenty of chances to use all that firepower.

I'd only been there a couple of months when we got a call about a stolen car that the Joliet police suspected was in Summit. I went looking for the vehicle and spotted the suspects at an intersection. At the time, I had a loudspeaker in my squad car. I'd sneak up behind somebody and turn that sucker on, and *boom*, guys just shit. In this case I got on that horn and told the suspects to pull over, but the sons of bitches took off instead, heading east out of Summit toward Chicago. I called for backup and took off after them.

Those guys had no idea they were messing with somebody who drove a hundred miles an hour every weekend at the racetrack.

Pretty soon we were doing ninety. They tried like hell to lose me, and when that didn't work, they leaned out the window and opened fire. When they turned down an alley, they lost control and hit a telephone pole. Three guys bailed out. One of them was firing at me with a sawed-off shotgun, but I wasn't afraid. I wasn't even ducking. I pulled over and jumped out and started returning their fire over the hood of my car. The gunpowder was so thick you could taste it. It was like a scene out of the movies. Finally my backup arrived and we were able to apprehend them. They ended up doing time for attempted murder. Me, I got a reputation as a wild man.

⊕ ⊕ ⊕

I hadn't felt that alive since Roger Douglas and I were in that car chase with the cops—and that had been years ago. I hadn't seen much of Roger. I knew he'd been making his living as a thief, but not much more than that.

Anybody who stole for a living in Chicago knew the drill. You paid the mob a tribute, usually through Willie "Potatoes" Daddano or Sam "Mad Dog" DeStefano—or you didn't live long. For some unknown reason, Roger had been ignoring that convention, which meant he was walking a very dangerous tightrope. But at the time I didn't know that—not until I walked into the station one morning and picked up the morning paper. The headline read "COPS FOIL BURGLARY ATTEMPT." According to the article, the Chicago cops—under the direction of Chicago's chief of detectives and rising star, Bill Hanhardt—had broken up a robbery in progress at a local liquor store. Three men had been apprehended. One had been shot three times, once in the stomach. That man's name was Roger Douglas.

Nobody seemed to know what happened after that. Or if Roger was alive or dead. I started calling around and finally found him, alive, in a dump of a hospital, Von Satzberg. It was an awful place, real old and out-of-date. People said you went there to die, not get well. I got in my squad car and drove right over.

When I walked into his hospital room, Roger smiled and said hello, but I could tell he was in terrible pain. I sat there on the edge of his bed and we talked for a while, just like old times. And then, between gritted teeth, he told me there was something I should know, "just in case." He told me that the deal at the liquor store was a setup. "It wasn't a fuckin' robbery in progress," he said. "It was a fuckin' ambush. That motherfucker Hanhardt ambushed us."

In 1968, Bill Hanhardt was the darling of the press, a one-man army and crime-fighting hero. But I'd heard he was as dirty as they come. If there were some thieves Willie "Potatoes" Daddano and Sam "Mad Dog" DeStefano wanted out of the way—maybe they were competition for the Outfit burglary rings, or maybe they weren't paying their tribute—Hanhardt would get a tip. A few days later, he and his coppers would happen upon a "crime in progress" and apprehend the burglars. Or if the Outfit wanted to get rid of the guys altogether, Hanhardt would stage a shoot-out. End of burglars, end of story. And one more rung up the ladder for Bill Hanhardt. It would be thirty-three years before the public learned the truth about Hanhardt. At the time, no one would have believed Roger—except maybe me.

"I'm smarter than that, Mikey," he said. "I've never been into stickups, you know that. And that's just where that son of a bitch Hanhardt's fucked up, too. He stepped on his dick this time all right. He figured we'd be armed, so when they gunned us down nobody'd think a thing about it. They'd just make out like they had to defend themselves. But we weren't armed. So they made up this big fuckin' story about how I picked up a hammer and they thought I was drawing a gun. But that's a goddamned lie. I was walking *out* of that liquor store with my fucking hands up. Yeah, I was giving myself up when those cocksuckers started shooting. Can you believe it, Mikey? I was completely defenseless. Fuck, man, I'm lucky to be alive. Three rounds, and one of 'em in the gut, too. When I get out of here, I'm gonna see to it they pay for this bullshit."

The whole time Roger was talking, he was looking worse and worse, pale as his bedsheets. I finally told him I'd better get going. "See you tomorrow," I said and headed out the door, but there wasn't a tomorrow—not for Roger Douglas. He died that night.

Three years later Bill Hanhardt was promoted to commander of the Burglary Division. The public idolized him. Me, I hated his guts.

As you can imagine, I was happy to see William Hanhardt finally get his: in October 2000 he was pulled out of retirement and indicted for his role as head of a multimillion-dollar jewel-thief ring with ties to Las Vegas and Chicago's infamous Tony Spilotro.

⊕ ⊕ ⊕

My career as a cop had been going great guns in Summit, but with Roger's death and the circumstances around it, there was a black cloud over everything. For a while I was having second thoughts about being a cop, but as I fell back into the routine, and Roger faded away, I realized I couldn't leave. Since I'd gone to Summit, Annette and I had been living very well. We had two brand-new cars and a nice apartment in a very respectable neighborhood. Annette had gotten a sitter for the baby and started working as a secretary for one of the VPs at Sun Chemical. We probably looked like the perfect couple, but it was no good and we both knew it. I was very unhappy in my personal life. If I wasn't working, racing, or screwing another woman, I was sitting at the Keen.

After more than a dozen near tragedies, the Keen's owner, Al Lorenz, had finally gotten smart and taken the bear and lion off the stage. So now it was just his girls up there, dancing in their white patent leather go-go boots to the latest rock and roll. I barely paid attention. But that night, when they started playing "Satisfaction" by the Rolling Stones, there she was. All legs and hips and breasts. She was beautiful, too. I took one look and that was it. Sex was all over this broad. I had to have her. A barmaid came by and I pointed to the stage and yelled above the music, "What's her name?"

I couldn't take my eyes off the girl they called "Candy Kane." She looked at me the whole time she was taking her clothes off. I wanted her in the worst way. I'd never experienced anything like it. After she got off work, she let me drive her home. We made love that night—if that's what you want to call it. Making love to Candy was wild. She wrapped those long legs around me and man, I was in another world. It was the most intense sex I'd ever had.

There was no one like Candy. After that first night with her, I felt like a teenager again. I played that Rolling Stones song probably a thousand times. Over and over. I couldn't think of anything but her. But what *was* I thinking? Here I was, a twenty-three-year-old cop, in love with a girl who worked part-time as a stripper. Was I nuts or what?

I told Annette it was over. Fortunately, our breakup was amiable; we split everything. I paid sixty dollars a week for support and continued to see Keith on the weekends. I couldn't wait for the divorce to be final before I moved out; I had to be with Candy. I got a nice three-bedroom apartment, and she and her two kids moved in. For a while I was in heaven. I'd never been in love like I was with her. It was probably one of the worst moves of my life. Candy was just a kid, nineteen years old, and very immature and jealous. She was also a total maniac. If she got pissed off, she'd start in with her games. We're talking the Olympics here; that woman was a gold medal winner. She could push all the right buttons.

I found out very early in our relationship that Candy got off on danger. Generally, I didn't take anyone out with me in Summit. Things were much more unpredictable there than they'd been in Willow Springs. One minute you could be eating a jelly doughnut and the next you'd be doing a hundred and twenty down the interstate with guys shooting out their windows at you. But Candy wanted to go with me on patrol. What was I going to do, say no to her? If I didn't let her come along, I knew I'd pay for it later; she'd pout for a week, or I'd stop by the Keen and see her making eyes at another customer.

So I let her ride with me in the squad car. And more than once I was involved in a high-speed chase with her laying on the front floorboard. On several occasions I was called to a crime scene while she was in the car and I'd have to call for backup because I couldn't transport the perpetrator, which caused more than a few problems with the department. But at the time, I thought it was worth it; the danger made her hot. After a chase or an arrest, we'd make love and it was incredible.

Whatever it was Candy and I had going, it was totally unbelievable. And both of us always wanted more. It was like a sickness. I was addicted; all I wanted was Candy. While things hadn't worked out with Annette, I was the type of guy who liked the idea of being married. I was raised that if you were in love, you got married. I was in love with Candy, so I figured why not? Next thing you know, I asked her to marry me.

We must've talked about tying the knot a hundred times while we were together, but something always came up to ruin things. We'd get in a major fight, and that would be the end of it until the next time I got sentimental and asked her again. We even had a baby together—Mark—and we still couldn't manage to get married. Partly it was me. Partly it was her. But mostly it was us. We were like oil and water. We were doomed.

CHAPTER 17

I was at the Keen constantly. I'd be out on patrol, and I couldn't stand it—I had to go by and see Candy. I'd never been the jealous type before, but now I was in knots all the time. She knew that, she knew the effect she had on men and what she did to me in particular. And she got a kick out of seeing me squirm. But it didn't matter that she got off driving me nuts, I still couldn't think of anything else but her. I had it bad.

Overnight, my life was turned upside down. Things were getting totally out of hand, especially with her emotional manipulation. I was lucky I didn't go to jail over her. I got so mad once that summer that I almost killed her. I was napping on the couch and happened to wake up and hear her on the phone talking to one of her girlfriends. When I heard her say she'd been doing her little "phone sex" routine with another guy, I lost it. I grabbed my gun and walked into the kitchen where she was talking. When she saw me with a gun, she dropped the phone and took off out the door. Our apartment was on a lake, and it was a real beautiful place. Nice neighbors, very upscale. Of course, she didn't care what the neighbors thought. She started running around the lake, screaming her head off, and I was right behind her.

For the first time in my entire life, I was out of control. I was actually shooting at her—well, I wasn't actually shooting at *her*, I was shooting at her legs, trying to stop her. People were running out of their apartments. It was a real circus. I imagine I would've killed her if I'd caught her. But I didn't. She kept on running, yelling for help. And finally I gave up and headed back to the apartment. Somehow I managed to cool off and come to my senses.

Nobody in the neighborhood called the cops, so I considered myself lucky given the scene we'd made. I didn't need a domestic on my record. Five hours later, Candy walked in the door. I was calm by then. I figured we'd talk things over. But she had something else in mind. A few minutes after she came back, there was a knock on the door, and who's standing there in the hall but two cops. One of them was my buddy. They weren't acting very friendly. They were real serious. I said, "Hey, what's goin' on, guys?" And I heard Candy behind me, laughing. She thought calling the cops on me was the biggest joke in the world.

That was the only time in my entire life I've ever decked a woman. I didn't even think. I spun around and knocked her out cold. *Boom,* she was down. My friend just shook his head. Then he and his partner turned around and left. They knew I was going nuts dealing with Candy. As cops they'd seen enough to know she was dangerous.

That wasn't the end of my night with Candy. After she came to, we made up. And then we made love. It was fantastic. We were in love all over again, like nothing ever happened. That's how it was with us: me and Candy were on self-destruct.

⊕ ⊕ ⊕

Like I said, it wasn't any secret I was having some major problems with Candy. Everybody on the force knew. So when I was called one night to investigate the report of a Peeping Tom who just happened to be in our neighborhood—and I ended up killing the son of bitch—I guess it was only natural everybody would look at me and wonder. Overnight I was in the jackpot. The rumor was, I'd shot the guy because he was peeping in my girlfriend's place. Supposedly I'd caught him looking at her and gone crazy and killed him. But in reality the guy had been peeping in windows for almost a year. The Summit police were very aware of his activities. We'd been trying to apprehend him for some time.

The guy was one weird cookie. He dressed all in black and only came out when it rained. On one occasion he'd attempted to get in a bathroom window while a woman was in the shower. Over time he was getting bolder. He'd started trying to force his way into homes. People would actually see him looking in windows. The community was up in arms. We were constantly on the lookout for this guy. The chief wanted him real bad, so whenever it rained we were out there hunting for him. Which was what I'd been doing when I got a call that he'd been spotted at a particular address—which just happened to be five buildings down from where me and Candy lived.

I requested backup and headed to the scene. When I got there, it was raining so hard I could barely see anything. The people who'd made the call were waiting for me. One of them pointed toward some buildings and said, "The guy's down there. You can see him. He's standing on top of a car looking into an apartment." I was thinking my backup would show up any minute, so I told everybody to get inside and I headed for the guy.

Talk about balls. The guy had made no attempt to hide. He was standing on the hood of a car, looking in a window. He was so into his little peeping deal that he didn't even see me walk up. I got next to him and said, "What do you think you're doing?"

He took one look at me, saw I was a cop—and jumped on top of me. He started punching, but I wasn't about to let him get away. I was throwing those punches back at him. We went back and forth until finally we ended up in an alleyway. Meanwhile I was still looking for my backup.

This guy was real scared. He did not want to be taken in. He was fighting like hell. It was still pouring rain. And I was starting to get tired. My backup hadn't arrived, and I'd had enough for one night. I told him to put his hands up, that he was under arrest. But that didn't faze him; he just kept wrestling to get away. That's when I reached down for my gun—and what do you know—no gun. The son of a bitch had my gun and had it pointed at my chest.

This was life and death now, and I wasn't just going to lie down. We started fighting for the gun. It was a tremendous struggle. I finally got my hand on it, but he had the better grip and had control of it. I knew what I had to do. I didn't think twice about it. I always carried a second gun, a .45, behind my back. I pulled it up between us and fired one shot. Dead on. *Boom.* The guy flew five feet through the air and landed facedown.

When I turned him over, there was blood everywhere. I put handcuffs on him—which is what you're supposed to do in a situation like that—and called for an ambulance.

They worked on the guy all night at the hospital. I was just going off duty when we got a call that he had died. Hearing that, I was a complete wreck. I'd never been in a life-and-death struggle like that. I was also very upset about the fact that I hadn't had any backup. I figured if the other cops had shown up, the guy would've been alive.

For three days I stayed home. I didn't want to see anybody. I knew that I'd been in a fight for my own survival. I also knew I wasn't a cold-blooded murderer. But that didn't change the fact that a human being was dead—and that I'd killed him.

⊕ ⊕ ⊕

The shooting opened a real can of worms. There was a big investigation. Nobody was saying it out loud, but I knew they were all wondering if I had caught the guy looking at Candy.

It turned out the guy was an out of towner from Little Rock, Arkansas. He was the son of a very prominent Arkansas businessman who was involved in the Republican Party. Talk about bad luck. Even after we had a big investigation and it was ruled a justifiable shooting, the family tried to get me for violating his civil rights. But we had a ton of witnesses to the shooting, plus the fact that the kid had four or five priors in other states for the same thing. He'd been caught before, paid a one-hundred-dollar fine, and gone on his way.

I imagine his father pulled a bunch of strings, because the FBI also initiated a major investigation. It was a pretty hairy situation. I was taken off the street for about a month to work the radio and complaint desk. I was bored to death. I was also scared to death. I knew people were still saying I'd gone into a jealous rage and killed the guy. I was pretty sure the FBI had heard the same story.

The FBI must've investigated that shooting for a year, but when it was finally over, I was cleared. I was starting to feel like a cat, like I had nine lives.

⊕ ⊕ ⊕

The shooting only added to my rep as a wild man. Bad guys didn't want to screw with me. By the late sixties I was encountering more racial violence, which was a nationwide problem. When it came to situations involving Black Panther types, they didn't care what your rep was—they were taking you down. This wasn't about stealing a car or holding up some tavern. This was about freedom and their rights—at least that's how they looked at it. That made them very dangerous. And as a cop, a white cop, I was Enemy Number One.

Summit had always been a strange community as far as race relations were concerned. Everybody pretty much worked together at a corn products factory. They got along just like they had for years and years. Their kids went to school together. The old folks went to church together. There wasn't any turmoil. Things were pretty quiet until a Black Panther group set up their headquarters in the community and started preaching hatred and separation. Then, sure enough, we started having some major problems.

One of the first things was a race riot at the high school. We got a call that kids were trapped in their classrooms, that they were being held as hostages at gunpoint. When we got on the scene, the situation was extremely serious. We learned that there were black kids inside with knives and baseball bats. Some were Black Panthers. They were working over some of the white kids pretty good. The poor kids were trapped and terrified, and everyone felt helpless to do anything to save them, because the mayor had refused to let us go in. There was concern that we didn't have the necessary backup to put down a real honest-to-god race riot.

We could hear kids screaming inside the school. We had no idea what was happening in there. It was terrible. I couldn't just stand by and watch people get killed. I told the chief, "Those kids need us. That's where we belong. Somebody's gotta do something, so it might as well be me. I'm goin' in."

I took four other coppers with me. We went down the halls, classroom to classroom, not knowing what to expect. On the one hand I dreaded opening a door, on the other I just wanted to blow the hinges off. It was very tense. We finally took the place and made about ten arrests of Black Panthers. Four or five ambulances took over a dozen injured kids to area hospitals. The papers said that Summit's cops had saved the day, so naturally, I didn't get fired. Parents thought we were heroes.

From that point on, racial tension increased in the area. We started having lots of firebombings. And there were snipers, too—men actually sat on rooftops, shooting at people. About that time we got a new mayor. He wanted the cops to be more aggressive, so we started doing surveillance on the Panther headquarters.

One night we got a tip that the Panthers were headed down Archer Avenue with a car full of guns. We managed to find them and pull them over. There were four radical types in that car, all dressed in military fatigues, with a half-dozen semiautomatic rifles and four or five pistols. As it turned out, one of them was the leader of the Black Panthers, Fred Hampton. The arrest got a ton of publicity.

I guess I shouldn't have been surprised when somebody wanted a little revenge. It was around two in the morning when we got a call about a suspicious car parked in an alley behind a tavern. But when I pulled into the alley and saw a car with two black guys, I immediately knew it was a setup. I looked in my rearview mirror—and whoa—another car had pulled in behind us, blocking us in. They had us like rats in a trap. My partner nearly shit his pants. I told him to get ready. And then they opened fire.

The first shot took out the passenger window and nailed my partner. Lucky for him it was nothing but bird shot. But he didn't realize that; he went nuts and started yelling, "I'm hit, man, I'm hit." We were getting some heavy fire. They took out the windshield. There was glass everywhere. I told my partner we had to get out of that car pronto. On the count of three I said, "Now," and he rolled out the door and under the car. I pulled a shotgun out from under the seat and rolled out the other side. I only had five rounds in the son of a bitch. I had my sidearm, too, but against those guys that was nothing. I managed to reach in and get the microphone. I called for backup, but you could barely hear above the gunfire. I said we had a man down, that we were pinned down. And then I waited.

We couldn't hold out much longer under the barrage. If our backup didn't show up soon, those bastards were going to waltz on over and pop us both. I had to do something. I got up to the front fender and let off three or four rounds. All of a sudden I heard shots hitting the side of the car—right next to me. I looked up, and what do I see but the tavern owner, an old Greek guy, shooting out the window. He's trying to help us. Only problem is, he's a terrible shot and he's firing at *us*. I yelled at him to hold his fire. He had no cover whatsoever. I figured those guys were going to take him down any second. But he wasn't listening to me—he just kept on firing out that window.

About that time our backup arrived and all hell broke loose. Before it was over there were probably a hundred cops on the scene. It was a tremendous gun battle. The entire deal was all over the radio and TV. They were saying that I'd been shot and taken to a hospital, but the truth was I'd barely gotten a scratch. I had some glass in my face, but it was nothing. Even my partner, who'd wailed like a stuck pig, didn't go to the hospital. As for the black guys, our dogs tracked them through to a building where we found some bloody clothes and another weapon, and on to the rail yard where one of them had bled to death. A few days later we arrested two suspects, but there was never any real conclusion to the incident. Everybody believed the ambush was retaliation for our arrest of Fred Hampton. We even had intelligence that said he was there that night, but we couldn't prove it. Later, after Fred Hampton was shot to death in his bed in what many in the press deemed a "highly questionable police raid on the Black Panther headquarters," every cop in Chicago was looking over his shoulder.

⊕ ⊕ ⊕

The racial and political unrest in America gave the mob the opportunity to flourish; while the public was watching the left hand, the right hand was busy screwing up the world. Certainly J. Edgar Hoover wasn't committed to busting up organized crime; instead, it was open season on the blacks, the hippies, the antiwar kids. Everybody blamed all of America's problems on them. Meanwhile it was the guy next door, driving around in a big caddy, who was really responsible for messing things up.

Fortunately, I wasn't at the Democratic Convention that year, but I heard that it wasn't the hippies and radicals who caused all the trouble in Chicago. Instead it was the cops and politicians behind those riots; they'd been looking for an excuse to beat up on the kids for a long time, and thanks to that convention, they had one. From then on, all kids saw was the badge, uniform, and gun. They were calling us pigs. It got so bad that I was ashamed to say I was a police officer. I even toyed around with the idea of leaving the force and going to Vegas. But I knew the high life of Glitter Gulch wasn't my style. I would've been bored to death, walking around in a suit, greeting a bunch of high rollers all day long. So Vegas was out. But believe it or not, Willow Springs was back in.

CHAPTER 18

In a year and a half I'd managed to make a hell of a name for myself in Summit, which absolutely fried Doc. I'd been involved in several major armed robbery arrests, where I went into the joints when they were being held up and took the guys down. I was the man. And I was nuttier than a fruitcake. Doc just hated hearing about how well I was doing in Summit. He called me every week to ask if I'd "seen the light" and was ready to come home.

I was having way too much fun in Summit to go back to Willow Springs, but I did miss Doc and all those stunts he pulled. I'd drop by the tavern and we'd chew the fat now and then. Over time he'd grown on me. He was a lowlife crooked bastard, the worst I'd ever seen—but he was also a goddamned Renaissance man. There was nothing Doc couldn't do—as long as he had a pocketful of money.

After almost two years, Doc had had enough. One day he got in his Ford and drove out to Summit to see the mayor. He told him, "You fire that son of a bitch Corbitt and I'll hire him back. Just don't you tell him I told you to fire him." And that's just what happened, too. It didn't take a genius to figure out who was behind it all, but I couldn't get too upset over Doc's end run. Besides giving me more money, Doc renewed his promise to make me chief after the next mayoral election, which was right around the corner. As far as I was concerned, being chief was the brass ring. It would've taken years to reach that level of power in Summit. I cleaned out my desk and moved back to the Willow Springs Police Department. And I did it with a smile on my face.

$$\oplus \quad \oplus \quad \oplus$$

I soon found out that the main reason Doc wanted me back was because he was having a hard time with the press and the county sheriff, a guy named Ogilvie, who later became governor. Ogilvie was a Republican and Doc was a Democrat. They were total enemies, mostly because Doc had all these patronage jobs on the side on top of being mayor, and Ogilvie hated that. He and his boys were giving Doc some real atomic heat. He'd even gotten the press on Doc.

But Doc was way ahead of Ogilvie; he was planning to use me as his high-profile crime-fighter to create a new image of Willow Springs law enforcement. Doc wasn't about to change his ways. He'd still be playing the same games. He'd just have me blocking the view.

I hadn't been there a week when Doc called and said, "I hear you wrote a ticket last night . . . an overweight truck out of Big Ben Chemicals? Well, tear the fucker up. You hear me, laddie? Tear it up. And you got the driver in lockup? Let him go. *Now.*"

Evidently Doc had gotten a call from somebody up the line about the overweight tank truck I'd pinched the night before. It was worth five or six grand in fines to the village. I got a piece of the deal and so did Doc. I tried to explain all that to him, as well as the fact that things had changed considerably in law enforcement: he couldn't handle business by throwing a ticket away. We had to be accountable. I also reminded him that the last thing he needed was more bad publicity, especially since he had an election coming up, an election I wanted him to win.

Doc yelled back, "I don't give a rat's ass about bad publicity. Tear that goddamned ticket up. Throw it plumb to Atlanta. I don't care what you do, just get rid of it. I want it gone. And I want that fella out of jail, too."

So, much as I hated to do it, I threw the ticket in the trash. But I didn't want to let that truck driver out of jail. When we'd pinched him, he'd been such a jerk that we had to arrest him and throw him in jail. When I went down to lockup to let him out, that little bastard grinned at me from ear to ear—like he knew he had me. It was all I could do not to kick his ass. Tearing up the ticket was one thing, but I just couldn't bring myself to open that cell door.

I decided to go talk to Doc face-to-face about the deal. He needed to understand that the guy deserved to be behind bars, that this wasn't just any pinch. Besides, we couldn't be doing this kind of thing every day. Eventually Doc would get caught, and I'd get caught, and he'd lose the election, and I'd never make chief.

When I got to the White Star Inn, I spotted an unmarked squad car. I thought I recognized it, and sure enough, sitting with Doc was this cop and known bagman Jack Clark. I'd never liked him. He was too showy, too pushy, too everything for my taste. On that particular day he was looking real flashy, wearing more gold than Mr. T. Doc waved me over, and right away Clark asked if I wanted a drink. I said no, that I was on duty—which the two of them found very amusing—and Clark put two envelopes on the table, one for Doc and one for me. "A little somethin' from the boss," he said and winked at Doc.

The boss? The boss was Mayor Richard J. Daley. He and Doc had been personal friends for years. Daley always stopped by Willow Springs to see Doc when he was on his way to play golf in Cog Hill. It was a you-scratch-my-back-I'll-scratch-yours deal, which was the way things had always gone between Chicago and Willow Springs. And whether I liked it or not, that's how they were going to stay. I smiled at Jack Clark and stuck that envelope in my coat pocket. Then I went back to the station and let that truck driver out of jail.

⊕ ⊕ ⊕

The mayoral election in Willow Springs was just a few months away when Doc started easing Joe Kresser out the door. The chief was in his late seventies, well past normal retirement age, so it was time for him to go. Doc knew that, but he also knew he had to take things slow because Kresser's wife was head of the Democratic Party. Pushing her husband out could create hard feelings, and Doc didn't want that.

But Joe Kresser wasn't just old, he was also one hundred percent incompetent as a police chief. He'd always been. The guy had no personality, no public relations skills, and no people skills. The only thing Joe Kresser knew how to do was hold his hand out. Doc knew that just as well as I did. Still, that wasn't going to make it any easier to get rid of the guy. The way I saw it, that was Doc's problem. But Kresser could tell something was going on—and that I was involved. Now if Doc needed something, he called me, not Kresser. Doc even had me scheduling the other officers. It was pretty obvious what was going on. Over the next few months, the tension between Kresser and me started to build. Finally we had a little conversation—if that's what you want to call it.

I'd gotten a call to go out to the preserve. When I got there— nothing. At first I figured it was a prank call, but then I started thinking about the ambush at Summit, and the thought crossed my mind that it might be a setup. I parked behind a thicket, turned off my lights, and sat there waiting. I had my gun drawn and my shotgun lying next to me. I was ready for whatever was coming down that dirt road.

I was getting ready to leave when Kresser showed up in an unmarked car, with another cop. That's when I got the picture. And I didn't like it, either. It had never crossed my mind that they might whack me. I went over to confront Kresser. First I told his partner to get out of the car. Then I got in next to Kresser and said, "Let's get somethin' straight, Joe. It's no fuckin' secret you've been makin' a killin' all these years. You've been rootin' every poor son of a bitch who comes along. Well, now . . . it's my turn. And there's not one fuckin' thing you can do about it . . . except move over."

Kresser got the message. Nobody was going to stand in Mike Corbitt's way. I got out and walked back to my car and that was the end of it. No more trouble from Kresser.

⊕ ⊕ ⊕

But there was plenty of trouble from other sources. I was very conscious of the fact that I'd been hired by Doc to help avoid problems, not create new ones. Even so, sometimes it seemed like in Willow Springs we couldn't stay out of our own way. Whenever I tried to do the right thing, something would go haywire. Like the night I got a call from another cop, a friend of mine named Lindsey, asking for help at the scene of a hit-and-run accident out in the county.

As it turned out, a kid that Lindsey knew was drunk and had been driving real crazy all over town. When he ran off the road and into another parked car, he panicked—bailed out and left his car. That was bad enough—that made it a hit and run—but the clincher was that he didn't just leave the scene, he left a car full of guns. So we're talking big trouble. And that was why Lindsey had called me.

At the time, I didn't know all that. I headed out to the scene with my sirens blaring, red lights flashing, doing a hundred miles an hour. When I got there, Lindsey filled me in. Since there hadn't been any witnesses, we cleaned out the kid's car and left it where it was. We were already back at the station having a cup of coffee before the county cops ever arrived on the scene. We figured we'd made it, that we were free and clear, but pretty soon here come those cops, screaming into the station.

Cook County had a sergeant in charge of vice named Jim Keating. I knew he did business with Alan Masters and that they were good friends. But I didn't actually know Keating, and I sure as hell wasn't going to trust him. When Keating walked in the door, he put it on real good. He strutted around, acting like a hard-ass. He was the same size I was but maybe ten years older. He had three other cops—big burly sons of bitches—right behind him.

Keating took one look at me and Lindsey and said, "I want to know who drives that squad car outside and I want to know now."

Lindsey and I looked at each other and tried not to laugh. We figured Keating had to be an idiot, coming in there like that. "I drive that car," I said. "So what's the problem?"

Keating looked me up and down. "We've got witnesses that saw that car out at the hit and run. So if you're saying it's yours, then I'm taking you in. Yeah, that's right. I'm arresting you—"

I cut him off. "I don't think you're gonna do much of anything except turn your ass around and walk out of here."

I stood up, and everything got real quiet. Keating's face flushed red and he took a step forward. His pals were there with him, but they were on my turf. This wasn't the county. They didn't have their reporters and cameras and their publicity-hungry sheriff to protect them. I looked over at Lindsey, and he stood up. And we went for those guys, *boom, boom, boom.* We beat the hell out of them, and then we took their guns away and threw them in a cell. We kept them there for twelve hours before we let them out.

That was my first encounter with the illustrious Mr. Keating. Unfortunately, it wasn't my last.

⊕ ⊕ ⊕

Like I said, there'd been a lot of outside pressure on our police department. I'd thought Ogilvie was bad—but when Joe Woods took over as county sheriff things got worse. He picked up the gauntlet from Ogilvie and started coming after us big-time. He raided us. He made lots of arrests. And naturally, he made sure to take the TV cameras and newspaper reporters along. Joe Woods liked to see himself on TV. He liked to brag to the reporters, "We're gonna close down all those vice dens in Willow Springs."

When the reporters came around to the tavern to talk to Doc, looking for his comment, it was a public relations nightmare. He'd be three sheets to wind, wearing some ragged shirt and pants with his bolo tie and cowboy hat. He'd look those reporters in the eye and say, "We don't have any crime here. None whatsoever. I don't give a rat's ass what the county sheriff says. Fuck Joe Woods. Fuck the county. We're incorporated in Willow Springs. And we're gonna do what we're gonna do out here."

The papers only printed the whitewashed version of what Doc Rust said, but that was bad enough. The thing was, Doc was telling the truth. We did whatever we wanted out there. We didn't close down a single joint. We didn't arrest anybody. But things just got worse. We never knew when those guys from the county were going to throw half the town's whores or gamblers in jail. Ordinarily we would have known ahead of time thanks to the county police radios in our squad cars. But now they weren't saying anything over the radio.

On the weekends, Willow Springs got wild, so we always beefed up. But if we got a mutual-aid call from another area, we'd drop everything in the village and go. Which was what we did one Friday night when we got a call from some county guys for backup out at a cemetery on the edge of town. But what did we find when we got there? Twenty county squad cars—and my old buddy Jim Keating. They actually drew down on us. They didn't arrest us, but they did handcuff us and haul us over to their headquarters in Bedford Park.

When we got to their station, a dozen TV cameras and reporters were waiting. Even though we hadn't been formally charged with any crime, they locked us up. They thought it was real hilarious. Once they had us all in jail, they raided just about every joint in Willow Springs. They knew they had to get us out of the way or we'd tip off the owners. It was common knowledge that Doc had installed a switch in the station that you'd pull and the lights would flash in all the joints in town. That was their warning to close everything down, that a raid was on the way. But thanks to Jim Keating, we never had a chance to let anybody know what was coming.

I sat in a cell by myself most of the night. I was cussing everybody I could think of, especially Keating. I had plans for that weasel. When he finally showed up, he strutted over and said, "You remember that beating I took, Corbitt? Well, maybe I'd better come in there and show you what a real beating's all about."

I had to laugh. The guy had a dozen cops standing by ready to help him, not exactly what I called brave. "Sure," I said. "Come on in . . . but if you think you're gonna come in here and beat the shit outta me, you're fuckin' nuts, buddy. Let's see you try it. Bring whoever the fuck you want. You don't have the balls to kick my ass . . . so come on, I don't give a fuck. Go for it."

Jim Keating never opened that cell door.

There was some bad blood between Willow Springs and the county after that incident, especially between me and Keating. I didn't like him and I didn't trust him. I also sensed that he wasn't like most other cops, that he was more than your average guy on the take. Eventually I'd find out what he was about. And when I did, I wouldn't just distrust Jim Keating. I'd despise him.

⊕ ⊕ ⊕

Cook County's raid on Willow Springs was all over the TV and newspapers. They figured they'd get some publicity and this Woods character would be a big hero. They got their publicity all right, but that stunt at the cemetery was the biggest mistake the Cook County Sheriff's Department ever made.

You didn't screw Doc—he could be dangerous. And very vindictive. He told me, "They think they had the last laugh. Bullshit. They ain't seen nothin' yet." Immediately, Doc got us off the county radio band and onto Bridgeview's network. Then we refused any mutual aid to the county. They called and called for help. They'd have a guy down, but we wouldn't assist them in any way. They could be dying in the road. "So what?" Doc said. "Fuck them."

It wasn't long before those county coppers came crawling back. They begged us to install a county radio as a backup in our cars. But Doc wouldn't have any part of it; he could really hold a grudge. "Joe Woods got too big for his britches," Doc said. He was familiar with Woods's background, he knew the deal on everybody. Doc told me that Woods had a sister who was politically powerful. She was personal secretary to President Nixon and later became famous for "accidentally" erasing eighteen minutes of Oval Office Watergate recordings. Doc called her the Republican broad with "paws and teeth." What Joe Woods's sister looked like or what she did in Washington didn't matter to Doc. He operated Willow Springs like it was a sovereign state.

Thanks to Keating and his crew, there were now plenty of people who wanted to clean house in Willow Springs. We were all under a lot of pressure. Doc was drinking more. He was also more forgetful.

And more violent. Now in his seventies, Doc was driving around town loaded and packing heat. He always carried a Bible in his car, but it wasn't for reading and praying. He had it cut out so he could hide his .45 inside. One time he ran a red light, broadsided a car, and actually shot the driver. Doc was already out of his car when the guy, who had a broken leg in a cast, went to get out. When the poor guy reached for his crutches, Doc thought he was going for a gun. He pulled out his .45 and shot him. Luckily, he didn't kill him. But still, it turned into a big deal. Doc was charged with discharging a weapon, and there was a threat of a lawsuit. The worst part was that the guy happened to be black. The press had a field day with that. But Doc handled that deal like everything else—he paid the guy off.

That winter Doc went through four new Fords. I tried to be at the tavern at closing so I could drive him home. He'd be so drunk he'd think it was a cab picking him up. He'd get out of the car and try to give me a twenty. I'd tell him, "I'm not a fuckin' cabdriver, Doc." But he wouldn't listen. He'd throw the twenty down on the seat and get out of the car. He wouldn't let me help him to the door, either. Some nights I'd watch him fall down three or four times before he got inside. When I'd try to lend him a hand, he'd try to fight me. "You think I'm old . . . I'm dying?" he'd yell. Sometimes, if I didn't get to the tavern before closing, I'd have to go looking for him. I just knew he'd be in trouble somewhere. And sure enough, I'd find his car all smashed up in some ditch. Doc never even got a scratch. He didn't care how many cars he tore up, either. He had a deal with Lake Ford, and the next day after a wreck, he'd call me up and say, "Laddie . . . come pick me up. I gotta go to Lake Ford again." And away we'd go.

I was getting concerned about Doc. I didn't think he was using very good judgment. I couldn't believe the stunts he was pulling. He always had his gun and his flask with him. He'd even started drinking in the village meetings. Believe it or not, he had me take him to a bunch of different furniture stores so he could find a chair that had a real high back, so that when he spun around in the middle of a meeting, nobody could see him take a shot from the flask.

During every previous mayoral election, Doc had stood at city hall all day long. It was the only place you could vote, and he'd stand there from the time the polls opened until they closed, giving everybody the evil eye. He'd have a fistful of hundred-dollar bills, and he'd stand there taking bets that he was going to win. There were always a few takers, and after he took their bet and they started to walk away, he'd yell out "Sucker" after them. When Doc won—and he always did—he'd be on their doorstep the very next morning, waiting to collect.

Part of the reason Doc could push people around was that folks were scared of him. They knew his reputation. They'd heard all the mobster and gangster stories. But fear alone couldn't explain why Doc had been so "lucky" every election day for nearly forty years. Like Chicago's Mayor Daley, Doc had a terrific political machine. Still, not everybody was going to vote for Doc. That year there was a movement to get rid of him. Some of Willow Springs's part-time cops were actually planning to vote against him. When Doc found out, he called them all in and fired them on the spot. Then he took their badges and threw them out the door.

⊕ ⊕ ⊕

With the election scheduled for April, which was right around the corner, Doc went into his reelection mode. He went door to door giving out twenty-dollar bills. It was his way of paying off the average citizen, and it always worked. He completely controlled the community. He controlled who got what job and what contract. He was very tight with Pat Marcy, the alderman from Chicago's First Ward, as well as Mayor Daley. Doc made sure the right people always got a piece of the pie, so guys like Daley and Marcy always took care of him.

Aside from Mayor Daley, Pat Marcy was one of the most powerful men in Chicago. And in the Outfit. That's not just my opinion. FBI agent Bill Roemer called Pat Marcy a "made man" when he testified before the Senate in 1983. Marcy might have been an elected official, but he was also a mobster, plain and simple. He was also Italian, having Americanized his name from Pasqualino Marchone.

Forget about mob guys like Tony Accardo and Joey Aiuppa, Doc called Marcy "the miracle man." Doc said that if you didn't get Marcy's blessing, you didn't get elected. Doc just loved Pat Marcy. And Marcy loved him.

Doc's contacts had always been invaluable. Between the corrupt elected officials and the known Outfit guys, he could get anything done, which was why Joe Ferriola—with Joey Aiuppa's blessing—was planning to move some big wire rooms into Willow Springs. This was going to be a huge Midwest wire service. The money was going to be tremendous. Ferriola brought in Sam "Wings" Carlisi to get things off the ground. They purchased a home on a secluded street for the operation and started checking on getting their phone lines. It was turning into a hell of an operation. And I was right there. I made a few inroads, just feeling things out, trying to get my foot in the door. Joe Ferriola and I were pretty tight, so I thought that was a real possibility.

Meanwhile, I started planning how I was going to handle things at the station once I was named Doc's new chief of police. I'd already decided the first thing I'd do was modernize the place, bring it into the twentieth century. I didn't think I was jumping the gun. Given Doc's election history, he was a shoo-in for mayor. If the votes didn't go the way Doc wanted, I knew he'd fix them somehow. When he didn't like the hand fate dealt him, Doc just cheated.

Doc might've been able to cheat his way through life, but when it came to death, he came up empty-handed. It was early one morning, just before the election, when I got the call. They said Doc Rust was dead, from "natural causes"—that and liquor and hard living. I was in shock. Doc had been my biggest supporter, and I didn't know what I'd do without him. Sure, he'd been a big pain in the ass, but what the hell, I'd come to love the old bastard. With him gone, my entire future had come to a screeching halt.

Doc's funeral was a very big deal. If it hadn't been so somber, it would have been more like a three-ring circus. I organized a huge procession. Everybody was there: Mayor Daley, Pat Marcy, and all the politicians, union officials, and cops. Thousands of local folks came to pay their respects. At the funeral home, I heard somebody say, "Doc Rust *was* Willow Springs." But that wasn't accurate. Doc Rust was *bigger* than Willow Springs. He'd made that two-bit town. Willow Springs was going to be nothing without him.

⊕ ⊕ ⊕

Even though Doc died right before the election, he was elected anyway. I'm sure he would've gotten a big kick out of that. We got an interim mayor and then had another election, and a guy named James Peters was elected. He was supposed to get rid of the corruption and clean up the town. Yeah, right.

With Peters in office, there went my shot at being chief of police. And as far as that wire service Joe Ferriola had been putting together with Sam Carlisi, that died on the vine, too. With Doc gone and "Mr. Clean" Peters in office, the Outfit wouldn't risk it. Despite Peters's attitude, I managed to get along with him in the beginning, but then he got a hard-on for Chief Kresser, and I made the stupid mistake of siding with Joe. So then Peters decided to fire both of us at a village board meeting. There were no grounds to fire me, so I told the mayor and village board that I wasn't going anywhere. Afterward, they put up a notice that we were fired, but that didn't stop me. I came into the station and continued to work anyway. Even though I didn't get paid, I kept working. Finally some of the trustees talked to Mayor Peters and got me back on the payroll.

James Peters tried to fire me a couple of times during his four-year term. He was a very honest man who actually believed he could clean up the town. But he had no backing at all—which you need if you make a promise like that. After one term, they ran him out of city hall. It seemed pretty clear the people of Willow Springs weren't ready for reform. But were they ready for Mike Corbitt? That's what I wanted to find out.

CHAPTER 19

In April 1973, Willow Springs got a new mayor, Walter Bucki. Bucki was a truly great guy. But don't get me wrong when I say that. He wasn't at all kinky. Bucki was straight, a former brigadier general. But I knew from the minute I met him that I could work with the guy, and I couldn't help but like him; he actually wanted to do the right thing. But to be perfectly honest, the first thing Walter Bucki did as mayor of Willow Springs was totally wrong: he appointed me chief of police.

Then again, Walter Bucki had promised me that. He owed me that. I'd practically bankrolled his entire campaign. Bucki wanted a real estate development to go through the village board and knew that under the old Peters regime, it wasn't going to happen. We struck a bargain: if Bucki won the election, he'd make me "boss"— which was what we all called the chief at the time—and in return I'd call on a few friends of mine to help his real estate deal along.

Still, my appointment wasn't official until the village board meeting in May. And when they voted, things didn't go so well. I didn't win their support. Things got pretty heated, and the village attorney had to step in and clear things up. He said the mayor didn't need the board's approval to appoint me. The election was just a formality, but it wasn't mandatory. I could see that Bucki and

I were going to have some real enemies if he went ahead and made me chief. But Walter Bucki was a man of his word. Over the board's objections he appointed me chief of police in Willow Springs.

The minute the word got out, I was hearing from all the guys. They knew that with me as chief, things would really move ahead. Everybody who was anybody was calling to congratulate me: Turk Torello and Joe Ferriola; Tony Orlando and Paul Payne; Alan Masters. Butch Blasi even dropped by the station to offer his support. He slapped me on the back and said, "Good work, kid . . . now go celebrate, have a fuckin' party." Then he handed me an envelope.

There was two grand inside. Talk about a party. I had everybody over to Kagels one night and we took over the joint. I felt like I'd finally gotten that break I'd been looking for. Things were looking up in my personal life, too. After some real tough times, it was finally over with me and Candy and I had a new girlfriend, Janice. Even my mother, who was still very skeptical about the corruption in Willow Springs, was pleased. And as far as Dad was concerned, me being chief was the greatest thing since cut grass.

It was a great party, maybe the best ever. But I didn't realize just how special that night would be until I went to pay the tab and pulled out the wad of dough I'd gotten from Butch. Inside was a handwritten note that said: "Remember what I told you . . . take care of your friends." It wasn't signed, but I knew it was from Sam.

Later that night I went by the station and sat out front in my squad car, all by myself. I took that note out and read it, and all of a sudden I got real choked up. It was like I was a kid again, standing in front of my Sunoco that first night—with the whole world laid out in front of me. I must've read that note a hundred times. I read it until I could almost see Sam's face with those dark eyes of his staring back at me. "Don't forget who your friends are," he'd said. Sam Giancana didn't have a thing to worry about. Mike Corbitt was his man.

⊕ ⊕ ⊕

The only crime we had in Willow Springs was what I created myself—or picked up from the old days. Like I said, Mayor Bucki was as straight as an arrow. The guy wouldn't take a dime, which meant that now, everything that Doc and Kresser used to do was rolled up into one big package: me.

Actually, one of the first things I did as chief of police was directly related to Doc and Kresser. They'd been selling badges for fifty years, and by the time I took over there were over six hundred stars floating around the community. Within two months we collected three hundred badges. Once I got them back in, I started a reissuing program. A star cost whatever you wanted to pay. I didn't tell guys that to get a badge would cost five hundred or a thousand bucks. They paid what they felt it was worth, and I was always very appreciative.

Besides reissuing the badges, I started dipping into areas nobody had touched before. For instance, we had all those strip joints and book joints and whorehouses that were in violation of the law—and it was like they were sacred; nobody ever bothered them. But I'd learned at Summit that there were times when it might be to my advantage to have a little raid. It could be very good publicity for the department. And for my image.

Relations were also improving with the county, so I sometimes worked with those guys on a raid. Maybe they'd need a hit to look like they were earning their keep. I'd pick out some joint in Willow Springs and they'd pinch them. A few days later the owner would come crying to me, wringing his hands. He'd say, "Jesus Christ, boss, what am I gonna do? Those county boys got me for gambling. Can't you help me out?" I'd have him call Alan Masters. We had it all worked out. Nice and neat. And in all those years, nobody ever figured out it was a setup. Those suckers thought I was a hero. They never knew I got a third of whatever Masters needed to make things right—which was always a nice chunk of change. Talk about sweet.

I was also concerned about my officers. All Doc and Kresser ever cared about was lining their own pockets. But I knew I couldn't hire sharp guys on what they used to dole out. I wanted the best cops we could afford. I never had a guy in my department walk into my office and say his kid was sick and he didn't know how he was going to make it on a cop's salary. If he needed money, I made sure he got it. I always took care of the guys on my force, which was how my Christmas Club came into being.

Christmas Club was a nice name for the biggest, best shakedown scam ever invented. It was probably my shining achievement as chief of police at Willow Springs—at least when it came to rooting.

Until I took over, nobody weighed trucks out there. In Summit they were always doing that; they'd weigh them and shake them down. I knew an opportunity when I saw one. The minute I became chief, I taught my guys the ropes. And I got thirty of the upper-echelon cops from all the other towns involved.

Every single trucking company in the city of Chicago had to "contribute" to our club or they'd get pinched when they drove through our towns. If they were overloaded, we confiscated their truck and threw the driver in jail. Before the trucking company could get the truck back, it had to put up the bond for the driver and pay the fine. It was cold-blooded extortion, that's all it was. We cleaned up. Eventually I had all the major trucking firms contributing to my club—something that gave me enormous clout in the transportation industry. Anything I wanted from those truckers, I got. It was like Christmas every day in Willow Springs.

⊕ ⊕ ⊕

Most of the trucking operations went all the way up the ladder to the top. In Chicago that meant, to some extent, organized crime and the union. Putting together my Christmas Club, I met some very influential guys, men like Louie Pike and Ray Sloshing, who were heavyweights from the Teamsters.

Originally I met Sloshing when we pinched him for drunken driving. He didn't say he was the chairman of the Joint Council of the Teamsters. He didn't even say he was with the Teamsters. Naturally, there was a major stink over this guy getting pinched. I got called downtown to the union and I had to make it right. After all, we had his arrest on record. Of course, I knew who to call to get that done: Alan Masters.

From then on, Ray Sloshing was my guy. He was a good friend to have because he went on to become the main guy on the Teamsters pension fund. He was also the head of the Central District Council, which covered the entire Midwest—including Ohio, Illinois, and Indiana, all the places where the Teamsters were a powerhouse. Years later, Sloshing went out west. I heard through the grapevine that he practically controlled Las Vegas, that he approved loans for companies like Bally Manufacturing and all those other organizations that needed loans to build hotels.

Between Sloshing and Louie Pike, there was nothing I couldn't get out of the union. And there was nothing I couldn't get done for other people, too. Which was how I came up with idea of starting my own business. I knew a guy who had a security company—Tri County Security—that had gone under. I bought the guy's license, with all the different phases you needed: a private investigator's license, a physical security license, and an electronics license. Having a security company was a conflict of interest for a cop. Even if it was on the up and up, it would look like I was out to shake down the area businessmen, like I was into the protection rackets. To get around any legal problems, I kept the original guy on the payroll as the license holder and paid him around a grand a month. Of course, he had no function whatsoever in the company.

I set up an office and got a secretary. Pretty soon I had a ton of business. Any major construction job that went on in Willow Springs, Countryside, Hodgkins, Justice, Hickory Hills, or Summit was going to need security, so if I saw that some major construction was starting up, I'd make it my business to approach the contractors. Usually I'd go in the squad car, in my uniform. But there was never any threatening. It was all very smooth.

Pretty soon my security company was doing some major business. And most of it came through the great contacts I'd made as a cop. I operated security at the Pritzkers' Hyatt Regency in Chicago, the Signature Building at 151 Wacker Drive. I also had a bunch of security contracts for hotels, groceries, and jewelry stores. Eventually I changed the company's name to Swift and sold it to a big company in Chicago for six hundred thousand dollars. Not bad.

⊕ ⊕ ⊕

Like Sam Giancana had said, I didn't forget my old Outfit pals when I became chief. I always took real good care of them. They set up this amazing gambling operation called the Willow Springs Ballroom. They'd have two or three hundred people in there playing cards and shooting craps. You had to be approved to get in, and it was all suits and ties. They had Las Vegas–style craps, blackjack tables, roulette wheels—all with professional dealers. The operators wore tuxedos. In the pit, they had stick men and cashiers. And they always had great live music with marquee names like Tommy Dorsey. There were gorgeous cocktail waitresses, tons of booze, food—everything. It was just like going to Las Vegas, only better, because you didn't have to leave town to get there.

On any given night, you might see Chuckie English, who was Sam Giancana's boyhood pal and a very big gambler; Joe Ferriola, who would drop by sometimes to shoot craps; Turk's crew—really, all the guys were out there all the time. Naturally, they never took a pinch. Ever. And neither did their other place, where they had a traveling game, in the old Willow Shopping Center.

It was nerve-racking taking care of that joint at the shopping center. The windows of the building were blacked out, and that helped some, but the parking lots had three, maybe four hundred cars out there at a time. No way could you hide something that big. Fortunately, the county police were getting their end. And of course, I was getting mine. So really, who was going to raid that joint?

⊕ ⊕ ⊕

More than anything I wanted to modernize the police department. The place had been a laughingstock since the first day I walked in the door and met Chief Joe Kresser. Fortunately, Mayor Bucki also saw the need for change in our law enforcement practices and pushed the village for raises in pay and better benefits. We spent more money on equipment for the police department than they'd spent in a hundred years. The station finally got its own radio room and its own radio operators with proper dispatch. We even got unmarked cars. Then we put in offices and had a judge's chambers built in the city hall. We actually started to hold court in our own village. And believe it or not, I even managed to hire a secretary.

Once I started reforming the department, I wasn't about to let anything stand in the way of making it the best around. If the village wouldn't go for something, I found the money myself. Law enforcement in Willow Springs was starting to get some respect. The word was out that you didn't want to fuck with a Willow Springs cop. I'd walk into a bar where there was a disturbance, and it was like I was Buford Pusser in the movie *Walking Tall*. I was always in plainclothes, with cowboy boots, and my gun was stuck in my pants or in my pocket. But they took one look at me and knew who I was—and it was all over. I mean, *over*.

⊕ ⊕ ⊕

There was nothing I hated more than a wife beater. And there was nothing I enjoyed more than bringing one of those sons of bitches down. In fact, I hadn't been chief very long when the opportunity to do just that presented itself. I was at home one night, eating dinner with my family—by that time I'd married Janice and adopted her daughter—when I heard on the police radio that this guy had shot his wife and was holding off fifty policemen. We'd arrested him several times before. He was a bum who beat his wife and kids. And he didn't just slap them, either. This guy beat the ever-loving crap out of them.

While I was hearing all this on the police radio, I also started hearing all these sirens going by my house. So I got my sergeant on the horn and asked him what was going on.

He said, "Yeah, boss, you heard right. The guy's shot his wife and now he's got us pinned down out here. He's shootin' out the window at us. He's got two or three guns, too. His wife and daughter got out a bedroom window, and we have 'em out of the house."

When I realized that there were no civilians inside that house, I went ballistic. "You're tellin' me there's nobody in that house but this son of a bitch and you guys are fuckin' around with him?" I yelled. "Evacuate the neighborhood," I said. "I'm on my way."

Then, believe it or not, I sat down and finished my supper. Nothing was going to happen until I got there, that was obvious.

I took my time getting to the crime scene, too. There were ambulances passing me. Fire trucks. Squad cars. But I didn't even break the speed limit. By the time I showed up, there must've been fifty squad cars parked out there. They had that place surrounded. Everybody in the county and the state was there. They had dogs, loudspeakers, chaplains. There were even helicopters overhead flying around with searchlights. It was a circus.

Every one of those guys had on a bullet-proof vest. And they were all huddled down in this big ditch between the road and the guy's property. It was just like my sergeant had said, the guy was shooting out the window. There were bullets whizzing by in all directions. Even with all his firepower, I still couldn't believe the way those cops were acting. You would have thought the guy was a SWAT team sharpshooter. I got out of my car and called my sergeant over. He came running, but he was crouching down, looking over his shoulder every second. He was shaking in his boots. I said, "Look at yourself, man . . . what the fuck are you doin'? Stand the hell up. You boys are wallowing around on the ground like fuckin' pussies. What the hell's goin' on here?"

He said, "The guy's shootin' out his window at us, boss. So we can't just stand up. We'd be sitting ducks. One of us could get hurt bad, sir."

I would've laughed if I hadn't been so pissed. "Forget about one of you gettin' hurt," I shouted. "What the fuck—you're here to protect the fuckin' public, the innocent people who live out here. If you happen to get hurt, well, so the fuck what? That's your fuckin' job. Now go get that bastard." He just looked at me, and I said, "*Now.*"

About that time, the guy started firing out the window again. My sergeant was shaking like a leaf. He shook his head. He started pleading, "Oh, no, boss. Please don't send anybody in there. This guy's got a gun. Shit, he's got three guns, maybe more. You don't understand, boss. He's dangerous."

I understood completely; we had a problem, and it wasn't the guy in that goddamned window. I said, "So you're saying that he's just going to keep right on shootin' at you and you're not going to shoot back? Is that the deal?"

"We're just waiting for some tear gas, boss."

By that time another one of my cops had come over. He was a young kid, a rookie. I'd liked him right off. He was very aggressive, very eager to please. I looked over at him and said, "You got a vest on, kid?"

He nodded.

"Good," I said. "Come on then, let's go."

"Go where, boss?" He looked at me real funny.

"In there," I said, and I pointed at the house.

That poor kid's eyes got as big as silver dollars. He said, "But we can't go in there."

I looked him dead in the eye. He knew I meant business. He shook his head. "I know you're the boss," he said, "but man, I can't go in there."

My sergeant nodded.

"I guess that means I'm going in by myself," I said and started walking toward the house.

I thought I knew that kid. I figured he was the type who couldn't let another guy walk into a dangerous situation like that all by himself. So I was betting he'd pull himself together and come through for me. Still, I was prepared to go it alone. But it was just like I thought, he was a stand-up guy. "Okay, boss," he whispered, "I'm with you." And away we went.

I'd already noticed it was real dark around the back of the house. So we went that way, across the grass and into the garage. We got down on our hands and knees. Our guns were drawn. I had a nine millimeter. We started crawling across that floor like we were military guys in combat. And all the time those shots were getting closer.

By the time we got to the door, the shots were on top of us. Our man was on the other side of that door. I stood up and motioned to the kid to do the same. "Okay," I whispered. I pointed my gun at the door. "Kick that motherfucker open."

So now I'm ready to nail that son of a bitch. I'm going to take him down. And what does that kid do? He shakes his head. The gunfire on the other side of that door was coming faster now. I wanted to use it as our cover when we went through the door. My heart was pounding. I hissed, "When I give a nod, kick the fuckin' door open." There was a shot, then another and another, and I gave him a nod, and *boom*—there I was, staring down a dark hallway at that beater. He had a rifle in his hand and he was looking out the window. He hadn't heard us. And he couldn't see us, because the hall and garage were pitch-dark. But we could see him. There was a light on in the house. I had him.

I thought his name was Jim, so I said, "Hey, Jim . . . it's over, buddy." He turned our way. He had that rifle pointed at me. He was squinting, trying to make us out, but I knew he still couldn't see me. I had some time. I said, "Jim . . . calm down. Listen to me. There's no choices. You're done, buddy. You and me are in here by ourselves now. You already shot your wife. You already beat your daughter. You already either threw 'em out the window or they jumped, I don't know. But that doesn't matter, because it's over. You've got to the count of ten to drop your weapon and get down on the floor."

I saw his finger trembling, hovering over the trigger.

"Don't do it, buddy. I'm tellin' you I'm gonna shoot you. You can't run, you can't move, because I got a bead on your head. Move and you're a fuckin' dead man. That's right, I'm gonna kill you."

He dropped that gun and fell to the floor. He was crying, acting crazy. We went in and got him and brought him out. We were heroes.

And that's just the way it went with me as chief. It was like I got to be bigger than life. I was always in the papers. My dad kept this big scrapbook.

There was one incident where there was a real screwed-up Vietnam veteran, a kid I'd known since he was just a baby. He was sitting at the kitchen table with a gun to his head, holding his mother hostage. Potentially, we're talking a very deadly situation. When I got to the scene we had five squad cars there, but nobody had gone in because of that gun the kid had pointed at his head. But me, I couldn't just stand there waiting for him to come to his senses—or blow his brains out—so I walked into that house just like I'd been invited.

When the kid saw me, he didn't take the gun from his head. There were big tears rolling down his face. His mother was sobbing. She was scared to death. I could see his finger on the trigger. And I knew this was the real deal. If I didn't talk him out of it, he was going to blast himself to kingdom come and leave his mother with a bloody mess. I used my most gentle voice. "How you doin', kid?" I said, and then, nice and easy, I walked over and sat down next to him. I talked to him for a long time—until finally he put the gun down.

In the paper they said I saved the guy's life. But the funny thing was, he ended up saving mine. As it turned out, he got his life back together and a few years later was sitting in a local bar when he overheard this broad trying to put a contract on me. He came to me with the information, and she ended up going to prison for trying to have me done.

$$\oplus \quad \oplus \quad \oplus$$

The Outfit was also keeping me busy. Tony Orlando and Paul Payne were still operating in Willow Springs. They were coming to me all the time, asking for favors, especially with some of the politicians I happened to know. I did whatever I could to give them a hand. They'd always been real good to me. There was always something extra for me in that envelope, and if a deal didn't work out, they'd drop something off for me just for trying.

After I became boss, I started to notice that Paul and Tony were getting very free with their dough. They'd always been generous, but this was more than that. They were getting a little forgetful, too, like maybe they were getting senile. Strange things started to happen. They asked me to help them move some machines into O'Hare. No gambling, just vending. So I did what I could. I called Doc's old pal and Chicago's First Ward alderman, Pat Marcy, to discuss Paul and Tony's operation. Marcy got them together with the head of airport security, but for some reason things didn't work out, which wasn't a big surprise. O'Hare was a real cherry for a machine operator.

So I figured that was the end of it. But then a few weeks later, Tony and Paul are coming around and they're the happiest guys alive. They'd managed to get into O'Hare. And for some reason they thought I was responsible. They went on about how the guys upstairs were very happy about what I did for them at O'Hare. As we were leaving Paul gave me an envelope and said, "This is for you and whoever else you gotta take care of. Every month you'll get the same thing."

You think I was going to give it back to him? Fuck no. I just thanked him real nice and put that envelope inside my coat pocket. When I got to my car and looked inside, there was five grand. I'd hit the jackpot. That was five on top of what I was getting for their regular operations in Willow Springs.

I'd been getting that little hit from Tony and Paul for almost a year when I got a call from Joe Ferriola asking me how things were going out there—were Tony and Paul treating me right, that sort of thing. At the time, Willow Springs was under Joe's jurisdiction, so the fact that he was keeping tabs on things out there wasn't any big deal. What *was* a big deal, what tipped me off that something was wrong, was what he asked next. Joe wanted to know what kind of numbers Tony and Paul were doing in Willow Springs. And he wanted to know what my end of the deal was.

In the last few years, Joe Ferriola and I had gotten pretty close, but I knew his real concern wasn't how I was being treated by those two old guys. What Joe wanted to know was if they were skimming off him. And with what he got out of me, he could backtrack and figure out what it was he should've been getting from Tony and Paul. While everybody skimmed, the trick was knowing how much was too much. I had to figure those guys had that part down pat after running machines all those years.

One day I was sitting in my office, and who do I see standing in my door but Tony Orlando and Paul Payne. I never saw those guys in a police station, so I knew there was a problem. They were real down in the dumps, too. They said I wouldn't be seeing them anymore, that they were retiring. Now they were both in their seventies and had more money than God, but they weren't ready to be put out to pasture. They wanted to keep hustling. But they said they didn't have a choice; there was new blood coming in. "Joe Ferriola's calling the shots out here now," Paul said.

A few weeks later, Tony Orlando and I met at Kagel's for lunch. He introduced me to their "replacement," Mike Pisola. Pisola, who I'd never seen before, proceeded to tell me that they had my envelope. "Everything's going to stay the same," he said.

I was relieved to hear that. I figured I was going to come out okay. Even if I ended up missing those two old guys being around all the time, I'd still make my nut. But then, a couple of days later, Pisola called up and said he wanted me to come over to the Four Es. He had a friend he wanted me to meet.

Naturally, I drove right over. I walked in and saw Mike Pisola sitting at a corner table. And next to him was Sal Bastone. I couldn't believe it. Sal was one of the guys who had stuck up the card game at my station, the cocksucker who almost got me killed. Right away Sal recognized me. He stood up. "*You're* the chief?" he said, like he was shocked.

"Yeah," I said. "You got a problem with that?"

He shook my hand. "No, I got no problem with that."

We sat down, and Sal started giving me the lay of the land. "Here's the deal," he said. "I'm the new guy out here. You know the position Turk and Joe had? Well, Joe's moved up and I'm that guy now. Mike here works for me. I'll come out here from time to time, we'll go out to eat. It'll be a good, long relationship."

I said that was fine by me, that I had every intention of going along with the program. We finished lunch, and Sal sent Mike out to the car. He wanted to talk to me alone. He said, "Okay, there's one more thing . . . you ain't gonna be gettin' that other envelope anymore."

"What envelope?"

A quick smile crossed his face. "You know the one," he said. "The big one. The one you shouldn't have been gettin' all along."

That lunch with Sal Bastone marked the end of my five grand a month. But I sensed it marked the beginning of something else— something big. Just talking to Sal I could feel it. Maybe that's why I didn't mind losing all that dough. Maybe I already knew that Sal Bastone and I were going to make one hell of a team.

CHAPTER 20

The Bastone brothers were golden. The holdup was ancient history thanks to having Joe Ferriola as a relative and Mad Dog DeStefano as their mentor, not to mention a couple of other aces they had up their sleeves: Sam Giancana and Hy Larner.

It didn't hurt that the Bastones also had some major brass balls. Old-timers saw them as tough—they liked young turks who were real scorchers. And the truth was, that type of guy could be very valuable in a certain capacity. The outfit liked to get ahold of a kid like that, the kind who always has to win and doesn't care how he does it or what it takes—whether it's lie, cheat, steal, or kill. This is a kid who has no conscience. And this is exactly the type of kid they want.

The old-timer will put him out on the street, working for him. He can go after a territory and be confident this kid will get it, and if somebody owes him money, it's pay up or else. There's no mercy. This kid doesn't screw around. He's not a nice guy.

The Bastone brothers were not nice guys. By 1974 three of them—Angelo, Carmen, and Sal—were made members of the Chicago Outfit. Of the brothers, Sal was the most ruthless and the least polished. At six feet, three hundred pounds, he was definitely not a front man. He never wore a tie and was so heavy he had to wear his shirt outside his pants.

The machine business—which was really the gambling business—didn't need social directors. It needed schemers who loved money more than they loved their own mother. Sal Bastone was perfect for the job. Some guys are into horses. Some guys are into casinos. Sal Bastone was into money. He loved gambling, but he never gambled. It was the business of gambling Sal got off on. Early in Sal's career, Mad Dog DeStefano had moved him and his brother Carmen into extortion. Sal turned out to be real good at strong-arming, and he became "the man." He was a mob banker, so to speak—a very dangerous banker. He put the dough out on the street, and then when the juice came due, he went after it. You paid or you got planted. Sal also collected street tax on porno joints and all the porno operators. No matter what it was, if it made a dime, Sal was there.

Sal and Carmen were already moving up the ladder when their big brother, Angelo, introduced them to Sam Giancana and Hy Larner. Angelo had been a state trooper until he got fired in the sixties during a scandal involving towing company shakedowns. He might have been finished as a cop, but as luck would have it, Angelo's life's work was just getting started. Sam Giancana took him under his wing and got him a spot at the Dunes in Vegas, working as Morris Shenker's bodyguard.

Morris Shenker was the mob attorney out of Saint Louis who represented Jimmy Hoffa. He was very connected to the Teamsters. In fact, it was Allen Dorfman, the union's pension fund manager, who loaned Shenker the money to invest in the Dunes and made him a very rich man. But not only was Shenker well connected to big money, he was also very well connected to some of the biggest names in the underworld, particularly the powerful Jews like Meyer Lansky and the Vegas newspaper mogul Hank Greenspun. Before long, Angelo had made quite a few valuable acquaintances in Las Vegas himself. He promoted his brothers whenever he got the chance, particularly to Larner and Giancana.

By 1974 the Bastones were considered an up-and-coming force

in Chicago. Angelo stayed out west, where he'd started looking after Hy's interests in Las Vegas. Sal was handling Hy's affairs in Chicago and the rest of the United States, particularly in Florida. Carmen was jetting all over the world, doing Hy's bidding in places as far-flung as Spain, Panama, Colombia, Iran, and the Philippines. When they'd first met Hy Larner in the late sixties, the Bastones were diamonds in the rough, not much class but a lot of muscle. Hy taught them finesse. He taught them how to curtail their harshness and how to approach people in a proper way, how to "speak softly and carry a big stick." But most of all he showed them how to keep a low profile. It was a very effective combination—Hy Larner's charm and financial brilliance coupled with the Bastones' cold-blooded ruthlessness. No one could deny that together they packed a real one-two punch. The question was, could they go the distance?

⊕ ⊕ ⊕

According to Pete Altieri, that punch came in real handy when Sam and Hy decided it was time to expand their operation. Pete had dropped by the station to shoot the breeze one afternoon—something he did on a pretty routine basis in those days. By the time I'd made chief, Pete had become very intrigued with Sal Bastone's crew. He said Sal was going places—fast.

Pete told me that since Eddie Vogel's retirement from the machine business, Hy Larner had been the power behind the scenes. Larner's latest conquest, thanks to Sal Bastone, was Zenith Vending. When Zenith first caught Larner's eye, it was owned by a Jewish guy named Kenny London. London was pals with Bill O'Donnell, owner of the Chicago machine maker Lion Manufacturing—or as it later came to be known, Bally Manufacturing. But as Pete reminded me, Bill O'Donnell was also very close to Hy Larner.

According to Pete, London had a big problem. He was a gambler, and by the early seventies he'd gotten in way over his head. It was looking like he'd take Zenith Vending down with him when O'Donnell and Larner decided that London had to go.

To make that happen, Larner reached out to Sal, sending his

most promising and most vicious muscle over to Zenith to pay Mr. London a visit. Sal didn't beat around the bush. He walked in and told London he was finished. From the way Pete described it, things got very tense. Sal was very threatening. Later Sal told me that he'd had to play hardball with London. He said he got right up in the guy's face and said, "Play it smart for once in your fuckin' life, Kenny. Pack up and get outta here. This is our business now. You fucked up, and now you're out." Sal said that even then London didn't back down. He started raising holy hell. He told Sal he didn't owe him a dime and tried to throw him out. But Sal was twice London's size. And very persuasive. Eventually Kenny London left Zenith and never looked back.

So that's how Sal Bastone and Hy Larner took over Zenith Vending. Pete said Zenith gave Sal his first real opportunity to demonstrate his business sense. Sal took Zenith from a losing proposition to a multimillion-dollar cigarette vending company. Then, using the revenue from Zenith, he started building up another company, Rentall Amusement. With Sal at the helm, Rentall was becoming extremely profitable. Supposedly, Sam and Hy were very pleased.

Of course, on paper those companies looked like duds. Nobody showed any money on their taxes because they were skimming off every bit of profit that was made on the cigarettes and what they called the kiddie games, like pinball machines. And what they were skimming wasn't just a handful of quarters. As I'd later learn, we're talking garbage bags full of money. Dozens and dozens of them, too. Every week. But it wasn't the kiddie games and cigarettes that were bringing in the big money. They were just a front for what Sal and Hy were putting together: a virtual gambling empire, built entirely of very sophisticated machines that were coming out of Lion.

If I'd thought the dough had been good back when I was at A&W, according to Pete it was nothing compared to what those guys were going to be doing in the future. He said the back rooms at Zenith and Rentall were full of all these new space-age machines. I hadn't seen Pete Altieri that pumped up about anything in a long time. For years, guys like Pete had made small fortunes in the machine business, but now it looked like Hy Larner was bringing new blood—and new technology—to the deal. Everything was changing. There was going to be a major shakeout in the business, no doubt about that. And if Zenith Vending was any indication, it was pretty obvious who was going to end up on top. Sal Bastone.

⊕ ⊕ ⊕

I went to Sal's father's funeral out of respect. I wasn't working with Sal on a day-to-day basis at that time. I hardly knew him. And I didn't know his father. All the big Outfit guys were there. It was quite a funeral, and Sal made it clear that he was very pleased to see me in attendance. He must've thanked me a hundred times. Going to that funeral was one of the smartest moves I ever made. It solidified my relationship with Sal. In the Outfit, paying your respects like I did that day can go a very long way. You're seen by a lot of people. It puts you in a certain circle.

After that, Sal started coming out to Willow Springs on a regular basis. He'd have his driver bring him to the police station and he'd come strolling into my office. He was totally fascinated by police work. He'd hang around for three or four hours at a time, wanting to see what was going on. He wanted to hear all my phone calls. And if there was someplace I needed to go in the squad car, he'd insist on coming along. He'd sit down, and then he wouldn't leave. Pretty soon it was like Sal Bastone was my deputy chief.

As I found out early in our friendship, Sal was a very inquisitive guy and very bright. After a few weeks of him coming by the station, I started suspecting that he had an agenda. He'd want to go out in my car, and once we got on the road, it was "Take me here, take me there." At the time, Sal's boss, Hy Larner, was operating out of Panama, so Sal was constantly wanting to stop and make a call to Central America. Or he'd want me to take him to the airport, usually O'Hare, to pick up a package or deliver one.

His ties to O'Hare had me curious. I couldn't believe how easy it was for Sal and his pals to bring things in and out of the country. I figured it was dough we were carting back and forth—although I knew it could have been dope. But nobody at the airport seemed to care what Sal was up to. It was pretty obvious Hy Larner had some very influential connections both inside and outside the United States.

The explanation I got from Sal wasn't very satisfying. "Hy's a fuckin' diplomat," Sal said. "The Panamanian government made him a diplomat for their country. So he's got a Panamanian government passport, a red passport. That means Hy Larner can do whatever the fuck he wants. He has diplomatic immunity. He knows everybody in a bunch of countries, including this one. He knows guys right up to the top, too. Shit, I bet we could transport nuclear weapons and nobody'd blink a fuckin' eye."

All that diplomatic stuff just made me more curious about Hy Larner. I couldn't imagine how a Jewish mob guy out of Chicago ever got so tight with a bunch of Panamanians. But then I reminded myself that stranger things than that were happening in Willow Springs. There I was, the goddamned chief of police, and I was chauffeuring this big-time mobster around town, picking up bags full of money. Now that's what you call strange.

⊕ ⊕ ⊕

Actually, Sal was a lot like me. That's why we got along so well. He was an all-day-and-all-night type of guy. He didn't ever want to sleep. He could go three or four days without a wink. And the guy was hustling twenty-four hours a day, making his rounds in the city. There were times he'd call and ask me to come pick him up at five in the morning. He lived way out in Deerfield, which was a long way from Willow Springs, so it was very inconvenient for me to go out there, but I did whatever he asked, no questions. I'd drive him around all day and drop him off at two the next morning.

Part of the reason Sal didn't sleep much was that he was very high-strung and fidgety. He didn't have any fingernails—he bit them down to the skin. He had to be constantly going here and there and taking care of business and making phone calls. He liked to talk, and when he wasn't talking, he was smoking or eating.

But with Sal it was all business. The man was not much fun at all. He didn't know how to party. It was almost impossible to get him in any kind of humorous mood, and he almost never laughed. In fairness, he did have a lot on his shoulders, a lot of responsibility to see that things got done right. Sal was running several huge machine operations—among them Rentall Amusement, Zenith Vending, and Apex Amusement. I got to see for myself what Pete Altieri had been talking about. They had all types of machines. Hundreds and hundreds of slots and so on. What today you'd call high-tech stuff. And of course, they had the traditional favorites— cigarettes, condoms, candy. "The three Cs," Sal called them.

We did a lot of traveling around the city so Sal could keep tabs on things. Fortunately, I didn't have a boss to answer to. But I didn't turn my back on the force. I'd made a commitment to Willow Springs. I'd started upgrading the police department and remodeling the Village Hall. I was hiring new police officers. I was calling in badges, shaking down truckers, and building my Christmas Club. I had a lot going on, but somehow I managed to hold everything together.

Everything, that is, except my personal life. I just couldn't seem to get that right. Leaving home before five in the morning and not getting back until after midnight didn't make for a good marital relationship. And it was a total mistake marrying Janice. It was a rebound deal, on the heels of my crazy relationship with Candy.

I'd been with Candy for several years and it had nearly killed me. I guess Janice, who was several years older than me, had seemed calmer, more normal. She was very mature and classy, too. In the time we were together, Candy totaled six Corvettes. New ones. Janice had her Vette washed and polished every week. It was immaculate. No matter how nice the place was that I got Candy and her kids, she trashed it. Janice had crystal chandeliers and a grand piano. Candy and I fought all the time. Janice and I never raised our voices. In fact, we didn't even talk. Unlike me and Candy, there weren't a lot of fireworks. There was just a lot of nothing. It was totally empty. Fortunately, I had other things to occupy my life besides my miserable marriage.

⊕ ⊕ ⊕

As I got to be more accepted by Sal and the guys around him, I started to be included in more confidential conversations. Sal had started testing the waters. He wanted to see how I handled myself. And Sal wasn't the only one watching. There were two other guys whose trust I was going to have to gain if I ever wanted to be a part of Sal's world. One of them was Bucky Ortenzi. The other was Hy Larner.

Bucky Ortenzi lived with his mother, a little old Italian lady who could really cook. You never went to Bucky's house that you didn't see sauces and gravies and macaroni all over the kitchen. Somehow, even with all that good food around every day, Bucky managed to stay trim. He was a health nut. He always ate fish because he'd read that fish oil would keep you young. He didn't work out, but he loved to walk. He'd walk all the way from his house in Melrose Park to the club where they had the bookmakers. Bucky was very close to Joey Aiuppa and Hy Larner. He was an extremely important member of the Chicago Outfit, but you wouldn't have known it by looking at him. He was very nondescript, with no flash at all. But he had tremendous power. He could've had a guy done in a second.

Bucky adored Sal. They'd talk for half an hour on the phone, and then, since Bucky didn't have a car, we'd go pick him up and have lunch at some Chinese joint. Sal got a real kick out of Bucky, especially the way he handled his money. Bucky was very frugal. Sal told me that one Christmas he'd taken him to Rocca Jewelers, where Hy had a credit line all set up so all the guys and their wives could pick out whatever they wanted. Some of them, like Tony Accardo, might walk out with a hundred grand or more in jewelry. Joey Aiuppa got a thirty-thousand-dollar watch one year. They'd bring their kids in, too. It was all on Hy. But what did Bucky Ortenzi get? "A little fuckin' tie tack," Sal said. He told me he turned to Bucky, while they were still standing at the counter, and asked him if he was nuts. But that was just Bucky. He was pleased as punch with that measly tie tack.

Maybe that's why all the Outfit guys trusted him with their money. Bucky Ortenzi was the counter, the sorter, the overseer. He was the keeper of the kingdom—that's what Bucky Ortenzi was. All the skim came through the doors of his little house in Melrose Park, dough from all over the world. At that time Chicago was skimming a million a week, documented by the FBI, out of the Desert Inn alone. The Outfit's total nut was in the billions.

The first time I met Bucky was at a Chinese restaurant. He was quiet and a lot older than I'd expected. Our relationship wasn't one of those we-hit-it-off-right-away deals. Bucky was very reserved. He didn't talk a lot. We had lunch at that Chinese joint a half dozen times before he started to warm up. I found out later he'd been assessing me the whole time, before he gave the nod to Sal. It was time to meet Hy Larner.

⊕ ⊕ ⊕

Hy Larner shook my hand and said, "I've been hearing good things about you, Mike." And then he went back to his dinner, and he and Sal went on with their conversation, like I'd always been there. Like they'd known me all their lives.

Over the next few weeks I began to know the mystery man the old-timers all called "Red." I picked up bits and pieces, little things about the guy. For instance, Hy Larner was much older than he looked—thanks to a nice tan and fine clothes. "Older than Methuselah," Sal said. "And richer than the fuckin' queen."

So there was the surface Hy Larner. The man with impeccable taste who wore thousand-dollar suits and five-hundred-dollar custom-made shoes and was never flashy. The man who was well-groomed and manicured.

Then there was the other Hy Larner. The private one. A quiet, soft-spoken man who deeply loved his Jewish heritage and had been married for years to a gal named Mert. Hy and Mert had several children; one of them, a son, had died of cancer. They had several homes, among them a million-dollar condo in Chicago and a penthouse suite in the hotel Hy owned jointly with Manuel Noriega, the Bambito, in Panama.

Hy had his own personal helicopter in Central America. A pilot came to the hotel in the morning and flew him into Panama City for his meetings with all the dignitaries. He also owned several other helicopters, a few jets, some props, and several cargo planes. He had a four-engine Fokker and a QueenAir—which is a big turbo-prop Beechcraft. And he had a KingAir, rated to about thirty thousand feet, that was used to haul freight. Sometimes his son Bruce, who was a commercial pilot, would fly the Fokker. But that was just for fun. Hy had six or seven million dollars in aircraft, so he always had a whole crew of pilots at his disposal. Hy loved to fly; he was constantly going in and out of the country, mostly from Miami and Fort Lauderdale, where he had homes.

I was almost as fascinated by Hy's wealth and power as I was by what he and Sal had to say about Sam Giancana. Other than knowing that Sam had been living in Mexico, I'd heard very little about him since he'd left Chicago almost a decade ago, back in 1966. From what they said, it was clear that Sam was still in the loop. He and Hy had teamed up all over the world. The men I'd met in Chicago—Bucky Ortenzi, Sal Bastone and his brother Carmen and his sidekick Joey DeVito—were just the small end of very large funnel. But in spite of all I'd heard, I still didn't have a handle on exactly what it was these guys were doing. Sure, machines made a lot of dough, but come on. This deal Hy had going with Sam was big, real big.

When Sal told me we were picking up Hy's wife, Mert, at the airport one day, I figured that maybe this was what they'd been sizing me up for, that maybe they'd wanted to make sure they could trust me as Mert's bodyguard and driver. Hy's wife wasn't exactly a kid—she was an elderly woman by that time—but all the same, she probably needed a little TLC.

Talk about having a deal all wrong. I took one look at Mert Larner and that big heavy metal suitcase she was lugging through the terminal, and I knew right away it wasn't Hy's old lady they were worried about. Mert could take care of herself. It was that metal suitcase they wanted me to protect. I picked up Mert dozens of times after that. She always flew into Chicago from Florida, and she never had any luggage. Except that carry-on metal suitcase.

⊕ ⊕ ⊕

Before long, I was getting involved with Sal's family. We were taking vacations together. We celebrated holidays together—Thanksgiving, Christmas, New Year's. Sal Bastone was my pal and was bringing me in on a lot of deals. He was starting to trust me. Of course, not everybody felt the same way. To some of the guys I was still a cop. After ten years with the Outfit, I still wasn't totally accepted by certain individuals in the organization. Joey "The Clown" Lombardo was the worst. He made a lot of derogatory comments about me, none of them funny. "The Clown" was not the perfect nickname for him, because Lombardo was not a comical type guy. He was linked to a bunch of murders. He would get to people I hung with and say, "Hey, this guy Corbitt's a fuckin' police officer. He's killed people. He's been in shootings. He's locked guys up. He'll fuck you, too." I heard about him grabbing guys out at Meo's restaurant and giving them a hard time about me, saying things like "What are you doin' with this guy Corbitt? Don't you know the guy's a friggin' cop? Don't even think about trustin' him."

Of all the Outfit guys, the only one I ever really despised was Joey Lombardo. Fortunately, he never said anything to me, to my face, that would have caused a confrontation. I would've loved to have whacked him. The guy was just asking for it. But I knew if I confronted him, he wasn't the type to take a stand. Lombardo was a coward, not a balls-out street fighter, the kind of guy who would've back-shot me or blown up my car. I knew his personality. If you got into any kind of argument with Joey Lombardo, the best thing you could do was finish him right there. If you didn't, it was only a matter of time; he'd take you out eventually.

So at that time, Joey Lombardo was one busy bee. He was running around trying to screw things up for me real good, complaining even to my friends. He actually went so far as to try to ruin things with Sal. But Sal would have none of it. He told Lombardo straight out, "Mike Corbitt's all right. He's a good guy. He's been working for Sam in Willow Springs for years. He's doing stuff for our crew now. He's a friend. You don't have to worry about the fuckin' guy. You're not sleeping with him. You doin' somethin' with him, then maybe you should worry. . . . Otherwise lay off him."

From what I understand, Lombardo even went to Joe Ferriola and tried to undermine my relationship with him. I'd just finished doing the security system at Ferriola's house—you don't give a job like that to just anybody. You've got to trust the guy you give that job to. And really, I guess that's what it all boiled down to. Trust.

⊕ ⊕ ⊕

In the Outfit, you don't give just any guy a million bucks and a plane ticket out of town. Trusting the wrong guy can get you killed. *Trust* isn't a word you'll find in a made guy's vocabulary. But for whatever reason, I'd passed the test with Sal and Hy and Bucky. I didn't know it yet, but I was about to join Hy Larner's elite group of couriers, or if you prefer the more common vernacular, I was going be a bagman.

They sent me to Las Vegas. On my first trip I wasn't carrying any dough. I was just supposed to make contact, do a little networking, that was it. "Just get the ball rolling," Sal said as he walked me to the plane.

When I landed in Vegas, some big thug guy I'd never seen before came up and said, "Hey, chief. We gotta car waitin' for you outside. We're supposed to take you to the Stardust." I knew I was supposed to go to the Star, but nobody had said anything about a car—or the three thick-necked thugs who grabbed my bags and escorted me to the silver limo with the dark-tinted windows.

So now it was my turn to trust.

I got in and we headed down to the Strip. For anybody who's never been to Las Vegas, it is really something. Even back then, when things were pretty basic compared to the over-the-top scheme of things today, the place was just a knockout. I got there around sunset, and I'd never seen anything like it. Cactus and hills and rocks and buildings—all silhouetted against this navy-blue sky. And this goddamned enormous ball of fire, all red and gold, blazing over it all. It was like the desert was on fire. And then, all of a sudden the sun goes down and the lights from the strip catch your eye. It's like you just ran into millions and millions of stars twinkling all around you, up and down that strip. I'd heard Las Vegas described as "Glitter Gulch," and I'm here to tell you, that's a terrible injustice. It's a beautiful place. And it's a strange place. But the bottom line is, there's nothing else in the world like it.

At the Stardust I checked in, and right away they started paging Frank Rosenthal. Although Chicago's Allen Glick was the hotel's president, Rosenthal—who most of the guys called "Lefty"—was really considered "the man." He was a Jew, so he wasn't a made guy, but that didn't seem to matter to the Outfit guys I knew. They all respected Lefty.

Fifi Buccieri, one of Sam Giancana's boyhood pals, had been Lefty's mentor in Chicago when he got his start in gambling. Over the years, Lefty had worked book joints, wires, you name it. He was good friends with my pals Turk Torello and Joe Ferriola. I'd also heard that he and Tony Spilotro—who had replaced Marshall Caifano as Chicago's liaison in Las Vegas—had gotten very tight.

Lefty showed up at the desk right away. He was a very striking man, a total gentleman who was extremely polished and immaculately dressed, all in pastels. Immediately he ushered me to a town house. There were only eight town houses, plush two-story jobs that opened right up to the pool. I could see this was going to be first class all the way. As he was leaving, Lefty handed me his card and said they wanted me to come to dinner later that night, that there were some people I was supposed to meet.

They had a fancy dining room all set up. Everybody was dressed up. A bunch of guys came in; mostly they were old codgers. One of them, Joey Aiuppa's brother, was a freelancer who just hung around at the Stardust. He had his own table in the Horseshoe Deli where he just answered pages and did favors for people all day long.

Nobody used their names at that party, which made it real

strange trying to figure out who was who. Nobody ever called me Mike or Mr. Corbitt. It was Mr. C. or chief. In order to get the lay of land, you had to watch everyone's body language, or pick things up through the grapevine later.

After a while some of the younger guys came in, among them Bobbie Stella, the casino manager. Bobbie was a Chicago guy from Cicero and was tight with Joey Aiuppa and Hy and that crowd. Besides working as casino manager, Stella was also a collection guy for Chicago, which meant he handled the skim—not that anybody said that. Everything those guys said that night was in riddles. Things were very veiled, so that nobody just came out and gave you someone's life story. But Bobbie Stella's reputation preceded him; I'd heard about him through Sal.

At the end of the evening, Bobbie and Lefty took me aside. Lefty said, "We want you to enjoy yourself while you're here. So relax. Have a good time." Bobbie told me, "You got a ten-thousand-dollar credit here, you got a ten-thousand-dollar credit at the Marina Club, and you got a ten-thousand-dollar credit at the Dunes. That should be enough fun for a few days. Anything you want, you just give us a call. Live it up."

And I did, too. I was still real crazy at the time. I spent their money. Since I didn't gamble, I bought tons of clothes. I loaded up my suitcases. Nobody ever said a word. I was pretty tentative, so I only spent three or four grand, but in my mind that was really something. Over the five days I was there, I went all over the Strip. On one of my last days there, I went to the Marina Club and walked up to the cashier's booth. I gave her my name, and bingo, it was like I was a king. She said, "You get up to ten thousand. How much do you want?" I could hardly believe it. I took fifteen hundred and went to a table and gambled it all away. I had to force myself to gamble like that; I just hated to lose that dough.

I met a lot of people while I was out there. Bob Maheu for one. Maheu was Howard Hughes's man in town and would later be implicated in a number of unsavory schemes, most of them revolving around the collaboration between the CIA and the mob in the assassination attempt on Castro. Maheu had a security company, and with that in common, we had a pretty good conversation.

From what I gathered, Maheu's operation was a lot more than your run-of-the-mill security deal, an assumption that was confirmed by Lefty Rosenthal on my last night in Vegas when I had dinner with him and his wife, Geri. Lefty told me on the side that "Bob's company [meaning Maheu's] gets all the broads for the government boys. All over the world. They do wires, all that type shit. He's very close with Sam and Hy."

Lefty and I didn't have much more serious conversation after that, thanks to his wife. Geri was a crazy showgirl who had a terrible temper and a mouth like a sailor. I listened to them fight and slur each other for the rest of the night. Lefty must have apologized fifty times. His wife never apologized once, but she was half in the bag when we started dinner. A blind and deaf man could've seen disaster ahead for Lefty with that woman. I wasn't surprised when I heard later about her fucking around on him with Tony Spilotro. The woman was a total slut. Beautiful, sure. But a slut just the same.

There were plenty of girls like Lefty's wife in Vegas and I had my share of them that week. It was all on comp, so I figured what the hell. It was probably the most amazing week of my life. So many doors were opening up.

Right before I left, I got a basket of fruit and a bottle of champagne. The note attached said, "Keep up the good work." It was from Sal. I might not have broken a sweat out there, but just making the right impression and putting together the necessary connections can be very difficult. From the sound of Sal's note, it seemed I'd hit a home run. He'd obviously liked what he heard from Lefty Rosenthal.

When I checked out, Lefty and Bobbie were there. They took care of everything. They walked me to the silver limo, and Lefty said, "Remember, chief, anytime you wanna come out west, you just give me a call. We'll take care of you." They shook my hand and that was that.

All the way back to Chicago I was flying higher than that

damned plane. It was all so new, so fantastic. I felt like a real big shot. I could see right away that I could get used to that way of life. I had a credit line of what—thirty grand? You would've thought I was a king. Ten years later I'd have a line of two hundred fifty thousand. Funny thing about it, by that time I'd feel like a pauper.

CHAPTER 21

Right after I got back to Chicago, things took off. I was involved in all the sit-downs and major conversations. I was going back and forth to Las Vegas once a month, sometimes more. I'd bring a bag stuffed with as much as five hundred grand back to Chicago every time, too. It was all skim. I was still a cop—at the station no one ever knew what I was up to, even when I went out of town—but now I was also a member of the Bastone crew.

"Hy is real pleased with how you're doin'," Sal said, which was a good thing, too, because I was beginning to see that when Hy Larner wasn't pleased, there could be hell to pay. Hy was a very soft-spoken guy, but like a lot of men with a quiet manner, you did not want to see him mad.

Around that time, there were several things Hy Larner wasn't very happy about—although I'd yet to see the full ramifications of that displeasure. Sal told me Hy was getting a lot of heat from the top-echelon men, guys like Joey Aiuppa and Tony Accardo, about the skim coming out of Vegas. Evidently it had been getting progressively smaller for months. There was only one explanation: somebody was dipping in the till. They hadn't fingered the man responsible yet, but Sal felt sure they would soon. I picked up from

Sal that there was a lot of suspicion around Tony Spilotro. Several of the guys thought he was a cowboy. Fortunately for him, most didn't, including Sam Giancana. So for the time being, the heat was off Tony.

I later found out that one of the reasons they'd decided to use me in Vegas was to see if there was a problem in that area of the operation. After I started making the trip, Hy began pulling couriers off the route one by one, to see if the envelope just happened to fatten up without that particular guy in the mix. When I finally found out they'd been watching me as well, I was not thrilled. Not that I was cheating those guys—I never did that—but I didn't like the idea of having big brother looking over my shoulder the whole time. It's a very uncomfortable feeling. Somebody says or does the wrong thing, and *boom,* now it's you that's in the jackpot.

Las Vegas wasn't the only place Hy Larner was seeing things slide. Locally there were problems as well. Over the past months, Hy's revenue from the independent machine guys had been slipping. He was becoming more and more certain that the independents were holding out on him. That spring he flew into Chicago from Panama specifically to talk about the situation and, over lunch, announced that he wanted Sal to go after "those greedy bastards." It was going to be an all-out push, a complete takeover of the other machine companies.

Sal told me later, "When Hy decides to go after somethin', he doesn't fuck around. These other cocksuckers better look out." The vending machine takeovers were a sign, like Pete had said, that things were changing in the Outfit. The power was shifting in Chicago. Sam Giancana was getting older, and a lot of the men from his 42 Gang days were now dead or dying. Years before, nobody would've dared move in on one of Willie "Potatoes" Daddano's joints; Willie would've whacked anybody who even thought about muscling in on him. But Willie was dead now, and his kid was running things, and he just didn't carry the same weight in the Outfit as old Willie had, which meant it was open season on all his joints.

Evidently Sam washed his hands of the whole affair, which left

Joey Aiuppa, who by this time was next in line after Sam, to give the blessing to go after Potatoes's kid. Aiuppa was a miserly old bastard. He was never satisfied; he always wanted more. I guess he figured that with Sal Bastone in charge of all the machine operations, he'd get a bigger piece than he'd been getting from Daddano's kid. He knew Hy would take better care of him, so he told him to go for it. All of it.

And that's exactly what Hy Larner did. Sal's crew busted the kid's operations wide open and scooped them up. After that got around, most of the independent guys fell in line pretty quick. They'd get a visit from Sal or Joey Lombardo or one of the other guys, and after a few "negotiations," they'd cave right on the spot and hand the whole deal over, lock, stock, and barrel. Usually they were compensated to some extent. Of course, it was never as much as what the operation was really worth. Maybe they'd get a few thousand for having the brains to step aside.

But there were exceptions. One springs to mind in particular, one guy who refused to move over. He was an old jukebox guy. He and his partner went all the way back to Eddie Vogel and Chuckie English. They had a small operation out in River Forest. They handled pinball machines, poker machines, pool tables—stuff like that.

Like they had with all the other small operations in the area, Sal's crew walked into their place and made their pitch. Forget about finesse, Sal went in and just flat-out told these guys, "You've been in this business thirty years. You've made millions of fuckin' dollars. But now it's over. You're out of business. We're taking over your operation. So how much do you want?" Sal didn't ask, "Hey, is your company for sale?" Nothing like that.

There's a way to get your point across to guys like Sal, and it isn't screaming "Motherfucker" at them and spitting in their face. But that's exactly what this old guy did. He was a real hardhead. He completely ignored the fact that his partner was back in the corner shaking in his boots, begging him to shut his big trap. That didn't faze him. He kept right on harping at Sal. He told him to get the fuck out. "You ever come back in here and I'll call the fuckin' police," he yelled. "Don't go near my locations. Don't touch my fuckin' equipment. Any of it gets broke and I'm goin' straight to the FBI." And then, believe it or not, he chased Sal right out of the joint, right out to the street.

That wasn't smart. After Sal gave me the blow by blow, he was still crazy mad over the deal. He said he'd wanted to choke the fucking life out that guy right there. But Sal was a businessman now; Hy wanted things done very low key, no heat. So it looked like Sal's hands were tied. A few weeks later the guy's partner showed up on Sal's doorstep. He said he wanted to sell, and he was begging Sal to take that company. He said he didn't have to answer to his partner anymore—he didn't have one. It seemed the guy was dead. He'd been found in his car in River Forest with his throat slit, stabbed at least five times. "Killer Unknown," the police report said.

At that time River Forest was a totally mob-controlled town, so the partner knew the crime wasn't going to be investigated. He also knew he could be next. Hy got that company. And when word got out about the hit on the old guy in River Forest, things got a lot easier. Sal and his enforcer pals, like Joey Lombardo, would make the rounds. And one by one they all fell. There was nothing to it.

One of the only companies Hy Larner didn't go after at that time was American Vending, which was the third largest in the city. It was owned by Louie "The Mooch" Eboli. Louie was a cold-blooded killer himself. And he was very close with Tony Accardo. So they left Louie alone for the time being. But eventually Louie got old and weak and they took American Vending over, too, and it became a part of Zenith.

That old one-two punch had worked its magic again. It didn't take long before Larner and the Bastone crew had it all. And just like I'd always imagined, my friend Sal was going be there, right on top.

⊕ ⊕ ⊕

By 1975 I was in Florida on a pretty routine basis, thanks to my friendship with Sal Bastone. Besides Las Vegas, I was also going to Miami on business for Hy Larner. But most of the time I was down there for pleasure—although a vacation with the workaholic Sal Bastone could hardly be called fun. On one trip we flew into

Tampa—which was pretty unusual because we generally hit Fort Lauderdale, where there were always a lot of Outfit guys. But this time Hy came in, and he and Sal met with a "couple of guys" over in Ybor City. It was one of the few times I hadn't been along with those guys in a long time. That alone made me curious. Of course, I didn't say anything about it. But after that meeting, I started to pick up bits and pieces—things like there were some problems with Trafficante and Marcello, like they were squawking about their cut on one of the deals they had going with Sam.

Hy flew back to Panama, and Sal and I hung around the pool in Tampa for the next few days—which was about all Sal could take of a vacation. We'd had a few drinks, the sun was hot. We were both pretty relaxed. We were talking a little business, nothing heavy, when out of the blue Sal said, "That Santo's one ball-buster, let me tell you."

"How's that?" I asked.

"Well, really it's not Trafficante. Or that weasel Marcello either . . ." He shook his head. "It's all Sam, Mikey. I mean Sam just does not like to spread the wealth around. Man, you gotta do that. Now Hy's gotta get involved. There's problems with the deals overseas. But Sam could give a shit. Sam tells the guys to go scratch their fuckin' behinds."

"You'd better have the muscle to back you up when you start with that shit," I said. "Sam's out of the country. Who's gonna take care of guys like Santo?"

Sal shrugged. "Hey, what the fuck can I say? I mean it's not a good situation. Sam's puttin' Hy right in the fuckin' middle. And after all Hy's done for him, too."

I didn't know what to say to that. To my way of thinking, Sam had made Hy Larner. What Sal was saying sounded like the tail wagging the dog.

Sal went on, "Shit, Sam would still be rottin' in a fuckin' jail if it wasn't for Hy Larner."

I sat up in my lounge chair. "How's that?" I asked. I didn't usually challenge Sal Bastone. Hell, I never did. But this shit about Sam, well—it was like Sal was attacking my own father.

"You remember when Sam was held in contempt by Hanrahan and his boys?"

"Yeah. They let the grand jury expire. But I don't see—"

Sal cut me off. "You don't see shit," he said, swinging his heavy oiled body around to face me. He took off his sunglasses. "Listen, for Christ's sake. It was like this: Hy got Sam out of jail. Hy and Meyer Lansky and a guy out west named Hank Greenspun who runs the paper out there, along with this other Jew, Al Schwimmer. Hanrahan was going to call Sam back to the stand and make him go through the whole deal all over again. Sam would've been in contempt and gone right back in jail. But Sam was fuckin' dyin' in there. Hy knew that. Shit, everybody did. Hy figured he had to get Sam out."

I'd heard all sorts of stories about how Sam got out of jail. Things like the guys in the Justice Department caved. Stuff about Bobby Kennedy telling U.S. Attorney Hanrahan to back off. But according to Sal, the truth boiled down to money—a million dollars to be exact. The way Sal told it, Sam had offered a million-dollar reward to anyone who could get him out of jail. (William Brashler, in his biography of Sam Giancana, *The Don,* supports this assertion, but Brashler alleges that Sam offered one hundred thousand dollars, not a million.)

It was that million dollars, Sal said, that had given Hy Larner an idea. It just so happened that right around the same time that Sam was trying to get out of the Cook County Jail, Israel was having some major problems with its Arab neighbors—specifically Egypt, Syria, and Jordan. Coincidentally, Washington was having its own difficulties with Egypt's president Gamal Abdel-Nasser.

Since the early fifties, Nasser had been involved with a growing anti-British and antimonarchy movement. Egypt's 1952 military coup, which led to the deposition of King Ahmad Fouad II, was considered his brainchild. Nasser had always been a radical, but when he became president in 1954, his ideologies and vocal opposition to Middle Eastern monarchies, royals, and their ties to foreign interests became more than the subject of idle conversation among U.S. oil executives. Sal said that both the Outfit and the CIA saw the charismatic Nasser as a threat to U.S. policies and holdings in the Middle East as well as to their allies the Shah of Iran and the Saudi royal family. Even more of a concern was Nasser's ongoing nationalization of many, if not all, of the foreign interests in Egypt, including the Suez Canal.

Over the next decade Nasser's popularity grew among the Arab public, and to the dismay of U.S. investors in the Middle East, by the mid-sixties he'd become closely aligned with the Soviet Union, signing an arms deal with Czechoslovakia in preparation for what he saw as the coming inevitable clash with Israel.

As Nasser's power and influence spread, concerns escalated among U.S. investors. It was a situation that was very similar to the one the CIA and the mob had faced in Cuba with Fidel Castro and Batista. Like that situation, there was a tremendous amount at stake. By 1965, U.S. corporations stood to lose billions; organized crime was facing a similar loss. Sam and Hy alone were taking over a hundred and twenty-five thousand dollars a day out of one Tehran casino.

From what Sal said, President Johnson saw the wisdom in removing Nasser from the playing field, but thanks to the public's negative sentiment about U.S. involvement in Vietnam, his hands were tied. His administration couldn't launch a war on Egypt and its allies—but Israel could. They just needed the arms and money to defeat Nasser. And that's where the CIA came in—as well as Hy Larner and his Zionist associates. At the Outfit's Stardust in Las Vegas, Hy contacted the casino's favorite high roller and international arms dealer, Adnan Khashoggi. For a price, Khashoggi said, the weapons could be had. Next Hy suggested that Sam put up that million he'd offered—plus a few more—for the Jewish state's war effort against Nasser and his allies.

In Panama, the CIA already possessed the elements necessary for a covert operation of this magnitude: the cargo planes and landing strips, the military staging sites, as well as the shady banks that would be happy to hide any cash used to purchase weapons from Khashoggi. There was also Hy's Panamanian friend, the Israeli agent, Michael Harari. A member of the Mossad, Harari was Hy's frequent guest in Miami and Panama as well as in Chicago, where he often visited the Israeli embassy on Wacker Drive. Any smuggling or communication with the Israeli government could be easily accomplished through Harari.

According to Sal, President Johnson agreed with the scheme; secretly arming Israel and sending its forces after a mutual enemy looked like the perfect solution to everyone's problems. Thanks to Sam Giancana, Israel was given everything it needed to do the mob's and the CIA's dirty work, which included crushing the radicals in Egypt and hopefully eliminating Nasser in the process. In return, Sal said, Lyndon Johnson ordered Attorney General Nicholas Katzenbach to call off Ed Hanrahan in Chicago—the grand jury investigation into organized crime was to be terminated. Sam Giancana was off the hook. According to Sal, all he had to do was cough up a few million for Israel's next military adventure.

On May 31, 1966, Sam walked out of the Cook County Jail a free man. Exactly one year and five days later, on June 5, 1967, Israel launched the Six-Day War, attacking Jordan and the aligned states of Syria and Egypt. With so much firepower at their disposal, the Israeli forces totally overwhelmed their opponents, crushing Nasser and his allied forces in less than a week.

With over ten thousand Egyptians killed in battle and a loss of thousands of square miles of Arab lands to the Israelis, Nasser—as well as his plans for social reform in the Middle East—had suffered a fatal blow. Although he remained in power until his death in 1970, Nasser would never again be a force to be feared. For the time being, at least, the Shah of Iran and the Arab monarchies were safe, as were the Outfit's Middle Eastern fortunes.

I imagine everybody figured they'd gotten just what they wanted out of that deal with Israel. But the truth was, they'd gotten a lot more than they'd bargained for. They just didn't know it yet.

hatever Hy wants, Hy gets," Sal grumbled. And in this case, Hy wanted Sal to pick out a birthday present for Panamanian big shot General Omar Torrijos. As it turned out, Torrijos wasn't just any big shot with a few scrambled eggs on a uniform—he was the top dog. So naturally, not just any present would do. It had to be a new Colt Commander, a beautiful sidearm, complete with customized grips bearing the Panamanian military insignia. Hy had left nothing to chance, giving Sal very specific instructions, all the way down to the type of presentation case he wanted the gun in.

So now Sal and I were spending the day at a gun shop, looking at pistols. "Yeah," Sal complained, "while my fuckin' brother Carmen's layin' on some fuckin' beach in Spain, we're doin' this bullshit."

Lately, Hy had been grating on Sal. There were days when if I hadn't known better, I would've thought Sal was nothing but Hy Larner's gofer. Sal was beginning to resent it, too. Out of earshot, he'd make cracks about Hy being a Jew, a tightwad kike—slurs like that. A lot of the Italians did that, but lately Sal had been going even further. He'd actually begun talking about how he'd like to "whack the little hebe." And when he wasn't talking about it, I knew that his wheels were turning, that he was plotting and scheming. One day he made me take him to the library to look up poisons and ways to murder people so it looked like an accident. He never mentioned who it was he wanted to kill—but I had a pretty good idea who he was after.

Sal was feeling the pressure. He was in charge of the Chicago business and he was doing a hell of a job. He'd been extremely loyal to Hy Larner, had always done everything the guy had asked him to do. But it was like it was expected. It might not have been so bad if Hy had spread more of the wealth around. Sal took a huge cut out of the Chicago business, but as far as getting a piece of the pie in all the overseas business Hy had going, he'd never gotten so much as a taste.

Sal was starting to beef about that situation in particular. He figured Hy had purposely kept him in the dark about just how much dough he was raking in overseas. But the thing that really got Sal was that his brother Carmen had become Hy's fair-haired boy. Recently Sal had gotten wind of the fact that Carmen was dipping in the till, and that was killing him. He wanted to get his hands on some of that money.

While Carmen could go all over the world, living it up in places like sunny Spain and Panama, Sal stayed back in Chicago, freezing his nuts off. As Sal put it one frigid winter morning while we sat in the car with a box of Dunkin' Donuts and a cup of coffee, "Who knows how many billions my fuckin' brother's takin' out of Central America and Iran. And now he's starting on Colombia. Fuck. And what do I get outta the deal? I can't even get a fuckin' ticket to goddamned Disney World outta that cheap Jew bastard."

I shrugged. "Hey, so it's ice in winter with the guy. What can you do about it? I mean Hy's like that. He's very secretive, right? So forget about it." At the look on Sal's face, I had to go on, "Hey, maybe Carmen wet his beak a few times. But on a routine basis? Forget about it. Nobody ever gets their hands in Hy's dough. Hy keeps things real close. Shit, he never even gives anybody a count. Nobody knows what he's really taking out of those operations. It's not just you. Hy Larner's not accountable to anybody."

Sal mumbled and stuffed a doughnut in his mouth. He knew I was right. Hy Larner was like an island. He never gave anybody a count, and he did what he wanted. From what I'd heard about the operation in Panama, they put stacks and stacks of hundred-dollar bills together and threw them into bags. Hy knew how much was there to the penny. But he wasn't telling anybody. And the fact of the matter was—who was going to complain?

Who would dare say a fucking thing to Hy Larner? Not Sal. Hell no. He was a pussy when it came to Hy. Sal might run around dreaming up harebrained schemes to whack the guy—but to his face, forget about it. There was only one man I could think of who had the balls to take Hy Larner out. And I wasn't sure he still had the muscle.

Besides, I'd already seen for myself that it didn't matter who you were—if Hy Larner wanted you out, you were out. Even if your name was Sam Giancana.

⊕ ⊕ ⊕

"You know, kid," Sam Giancana said, "since I've been away, you've been gettin' way too much press." He tapped the front page of the newspaper and then shook his finger at me. "You can't be doin' this kind of shit. It's bad for everybody. . . . It puts other guys in the light. We don't want that."

Sam had been back in the States for almost a year, but I'd heard he hadn't come willingly. The previous July—1974—Mexican immigration agents had dragged him kicking and screaming out of his villa in Cuernavaca and flown him to San Antonio, Texas, where he was met by FBI agents and handed a subpoena to appear at a grand jury hearing in Chicago.

Given Sam's connections to U.S. and Mexican officials, I'd been surprised to hear of his deportation. But as I later discovered, although he might have had a number of friends in high places in the U.S. government, among them CIA agents, those contacts had virtually no knowledge of the day-to-day activities of a few gung-ho FBI agents; they'd been just as blindsided by the whole affair as Sam was.

From what I've been told, from the first day Sam took up residence in Cuernavaca, the FBI boys in Chicago had been pressuring the Mexican government to deport him. Thanks to Sam's friend the Mexican official Jorge Castillo—and thousands in payoffs—he'd managed to avoid the problem for several years. But for some unknown reason, something had gone wrong that past July; Jorge Castillo was conveniently "out of the country" when the Mexican immigration agents apprehended Sam. Without his patron there to protect him, Sam was deported and whisked back to Chicago.

Since returning to Chicago, he'd had a bunch of serious health problems. I heard he made the rounds with Butch Blasi, just like in the old days, only now it was more social than business. Occasionally he had a few of the old-timers over for dinner, and sometimes he met with the up-and-comers, young turks like his neighbor, Tony Spilotro, who came back to town from Las Vegas on a regular basis. When Sam greeted me that morning in his robe and pajamas, I immediately realized that all those meetings were just his way of staying in touch and nothing more than that.

I wasn't the same punk kid he'd last seen. I was grown up. A man. But I hadn't realized that he'd be different, too. That he wouldn't be the same guy I'd met at the gas station. The man sitting before me at the kitchen table now was aging and frail. He wasn't the old Outfit powerhouse who had swaggered around on TV. Bent over from pain, he grumbled about the recent abdominal surgery he'd had in Houston, not about the FBI agent he had tailing his ass.

For months I'd wanted to drop by to say hello. I was a chief of police now, and I figured he'd get a kick out of that. I hadn't expected he'd have newspaper clippings of my more infamous law enforcement exploits—or that he'd take issue with them. If I'd anticipated that, I probably wouldn't have jumped at the chance to deliver a package to him for Sal.

Sam glanced at the stack of newspaper clippings and said, "So what do you say, Chief Corbitt? How can we keep that gun of yours in its holster?"

"I don't know," I said with a smile and a shrug. "I mean, I *am* a cop. So I'm supposed to defend innocent people. If I have to take a bad guy down, well then, that's what I have to do."

He frowned at me. His voice was a growl now. "And then the papers pick it up, and there you have it. You're suddenly a hero. That's nice. Makes you feel good, I imagine. But I'm tellin' you, you shouldn't put yourself in that type spot. You have to stay out of the limelight."

I was surprised by his reaction. It was pretty clear he wasn't very happy with my success as a cop. Of course, making me into a big-name crime-fighter hadn't been the idea. Sam had just figured they'd have me over there in the middle of nowhere and if something came up and they needed a little help—there'd I'd be, sitting on my hands, waiting for them to call.

"That's not my style, Sam," I said. "I never ran away from a fight. And I don't stand by and let defenseless people get run over just so I don't get my name in the paper, either."

Sam gave me a sideways look. A smile crossed his face, and he started to laugh. "Okay, so what *is* your style then, Chief Corbitt? A big fancy car? I hear you're racing out at the speedway. A mansion out in the suburbs? A big boat?"

"Come on . . . you know I've always loved racing. Shit, I love cars. You know that. And as for a fuckin' mansion—I'm still living in an apartment. I've got about as much chance of getting a damned house as—"

He cut me off. "An apartment? Are you kidding me?"

I shook my head.

"But you're the chief of police now. You're not just some common rummy drivin' around in a car all day. You can't live like that."

I started to laugh. "Well, it's not like I have a choice. I've looked around, but you know things are expensive. With down payments and all."

Sam stood up and limped toward the hall. "You wait right there. I've got a little business to attend to."

I didn't have to wait very long. When Sam came back, he had a brown paper bag in his hand. He sat down and shoved it across the table. "Here," he said. "There's twenty-two grand in that bag. I think that should get you started."

"I can't take this, Sam," I said. "I mean, it would be forever before I got it paid back."

"This is a gift. You got that? You don't have to pay me back . . . ever. I want you to have it."

"I'd have to pay you back sometime," I said. "It wouldn't be right."

"Well, get out of that apartment and get yourself a proper house, and then we'll talk about it. Until then, forget it. In the meantime, do me a favor . . . stay out of the goddamned papers. You don't need the heat. We don't need the heat."

"Anything I can do for you, Sam, you just name it."

He grinned. "You better watch out, I just might someday, too." He sighed and rose from his chair.

I knew it was time to go.

At the door, he shook my hand. He still had a good handshake. "Now remember, I'm gonna want a full report on that new house you're getting, so I expect you to come back soon."

"I will," I said. But I didn't know I was lying, that I'd never see Sam Giancana again. Not alive anyway.

⊕ ⊕ ⊕

Early one morning just a few days after I met with Sam, I got a call from Sal. "Don't pick me up today," he said. He sounded real distant. Almost guarded. Now with anybody else I wouldn't have said a word, but this was Sal. I considered him a friend, so I was concerned. I had to ask if something was wrong.

"I'm fine." He sounded rushed. "Something's come up, that's all. I'll call you later. I gotta go now."

From the sound of Sal's voice, I knew something wasn't right. He'd been pretty open with me about most things, even when we talked on the phone. But this time I sensed he was being secretive. And that made very uneasy.

When I walked into the police station later that morning, I

found out what was wrong. A group of my officers were standing around talking. "Did you hear the news?" one of them asked.

I shook my head. "What news?"

"Sam Giancana got killed last night. Somebody went in his house and blew him away."

I went into my office and closed the door behind me. So that was what was wrong with Sal. He knew about Sam. I figured there was probably a big sit-down someplace.

When a boss gets taken out, guys disappear. Not that they always get whacked—they just run for cover; they go out of town and let the dust settle. If it was a mob hit, which was what this deal had to be, then anybody next to him could be in some major trouble. The Bastones had all been close to Sam Giancana, especially Angelo. I had to think his murder would really shake them up. And what about Hy Larner? He had to be shitting in his pants right about now. Everything had to be up in the air for those guys. Hell, they could all get whacked in the next twenty-four hours. That had been known to happen.

I didn't hear from Sal that day. Or the next. Needless to say, I was freaking out. I couldn't eat. I couldn't sleep. I just waited by the goddamned phone. Finally, three days later, Sal showed up at the station and said, "Let's go for a ride."

We got in my squad car, and the first thing I said was, "What the fuck happened?"

Sal wouldn't look me in the eye. He acted very uncomfortable. He just shrugged and said, "Well, I don't know for sure."

But he knew. I knew he knew. "Yeah, right," I said sarcastically. It was one of the few times I ever came back at Sal Bastone. But I thought I deserved more than that. After all, Sal knew how I felt about Sam Giancana.

Sal was quiet for a minute and then said, "Are you sayin' you don't have any idea who did this? Now don't fuck around with me, Mike. Come on. I mean there's only a few guys Sam would let in his house late at night like that."

"So you know who it was?"

He glared at me. "Hey . . . did I fuckin' say that? Did I fuckin' say I knew anything for sure? I'm not sayin' I know for sure. I'm just tellin' you what I think . . . that's all."

Obviously, the subject of Sam's murder had Sal very tightly wound. I wasn't going to push him any harder. Besides, sometimes in the Outfit there are things you don't really want to know. "Okay," I said. "So you don't know for sure."

"Yeah, that's right," he said, narrowing his heavy-lidded eyes. "It's just my own thoughts on the deal. . . ."

And then, like we'd totally changed the subject, like we'd never even been talking about a murder, Sal said, "You know, Sam sure loved that little guy in Oak Park . . . Tony Spilotro. Yeah, he was fuckin' crazy about him. Sam put Tony on the fuckin' map, thought he was gonna be a big fuckin' man someday. Did you know that after Marshall Caifano got out of Vegas, it was Sam who wanted Tony Spilotro out there? Even lately, with the problems out there with the skim and all, Sam always stood behind the guy. Tony was over to Sam's house all the time. He lived right by there. Did you know Tony even figured a way where he could get in through the back of Sam's place without anybody ever seeing him? He'd go through other people's yards, go over fences, all sorts of shit."

It was pretty obvious where Sal was headed. "Sam wouldn't open the door for just any son of a bitch," I said. "I mean there's Butch, Chuckie English . . . he'd let them in, all right, but shit, no way they'd ever do anything to hurt Sam. No way."

"Yeah, Sam and Butch were real close. And the same thing with him and Chuckie. Besides, neither one of them had the balls to do somethin' like that. There's only one guy that had the balls to do Sam."

Sal didn't have to say the name. I knew he was talking about Tony Spilotro. "But why?" I asked.

"There's never just one reason for shit like what happened to Sam. There's a million of 'em. Let's just say that maybe Sam should've remembered what happened to Bugsy Siegel."

I stopped at a light and looked over at Sal. His eyes were as cold as I'd ever seen them. I couldn't believe what he was saying. Bugsy

Siegel was the man who first envisioned Glitter Gulch as the Outfit's dream come true. Nobody saw the potential at first, not even Bugsy's New York boyhood pal Meyer Lansky. Bugsy made them all believers. And he made them all real rich, too. But at some point he started getting out of control, threatening the entire operation. Meyer Lansky saw it before anyone else. And when it was decided that Bugsy had to go, it was Meyer Lansky who pressed the other guys to have him whacked. Lansky might have been Bugsy's best friend and protégé, but he was a businessman first. Everything comes second to guys like that. There's no room for sentimental friendship with guys like Meyer Lansky—or Hy Larner.

I suddenly saw the parallel. But still, I could hardly believe what Sal was suggesting. "Are you saying that Hy—"

Sal cut me off and pointed to an Italian joint across the street. "Man, I feel like I haven't eaten in days. I'm fuckin' starvin'. What do you say we get a beef sandwich? And then I'll give Hy a call down in Panama."

It was like shutting off a switch. Sal and I didn't discuss Sam Giancana's murder for the rest of the day. In fact, we never did again. There'd be times when Sal was talkative enough—occasionally he'd make some offhand comment about how Sam had refused to share the wealth. But I never pursued it. Maybe I didn't want to know the truth. Or maybe I already did. Maybe I just didn't want to believe it.

⊕ ⊕ ⊕

During the following year, I picked up all sorts of rumors about Sam's death. There was talk about how somebody in the government had taken him out. Supposedly, one of Sam's own daughters claimed it was a CIA hit. And then there was the theory that it was a joint deal between the CIA and the mob, like the attempted hit on Fidel Castro back in the sixties—which was one of the revelations that came to light during Johnny Roselli's testimony to the Senate Select Committee on Intelligence in 1975.

The committee was investigating the U.S. government's involvement in a number of foreign assassination plots and political dirty tricks. As it turned out, almost all of them were collaborations between the Outfit and the CIA. Sam had been subpoenaed to testify before the committee. In fact, he'd been scheduled to leave for Washington the day following his murder. Some people thought his death was an extremely happy coincidence. As for me—I didn't know what to think anymore.

But a year after Sam's murder, when Johnny Roselli turned up floating in a barrel in Biscayne Bay, for some reason I wasn't a bit surprised.

CHAPTER 23

Nobody had more revenue from pinches than Willow Springs. We broke every record in Illinois. The only department that brought in more money was the city of Chicago. And they had fifteen thousand cops on the street. With so much dough to be made, it was no wonder cops from different departments got into conflicts. As time went on, things started getting out of hand and we became like rival gangs. The biggest problem was between Willow Springs and the county boys, particularly between me and Jim Keating, and particularly when it came to towing vehicles.

Tows were a huge moneymaker for everybody—except the owner of the car. A regular tow, where there was a breakdown or an accident, was not as profitable as the type where a guy got pinched for reckless driving or drunk driving. The cop called in the towing company. If he called a friend, he got a nice chunk of change for the referral—which, of course, is illegal as hell. A regular tow would bring ten, maybe fifteen bucks on the side. With the bigger offenses like a DWI, where you made an arrest, you'd get fifty or more for the tow *plus* whatever Alan Masters kicked back.

In Willow Springs, our bread and butter came from the regular tows. Which is why there was a constant battle between us and the county for them. We'd hear about an accident outside our jurisdiction, and we'd drive like hell to get there and do the report and get the cars towed away before anybody ever notified the county. That didn't make the county boys very happy; it was like you were stealing the dough right out of their pockets. I just loved sticking it to Jim Keating every chance I got.

There were several occasions when we were outside our jurisdiction and the county showed up while we were in the process of having a car towed away. It always ended up in a fight—just to get that tow. It got so bad that we carried maps in our squad cars showing where our jurisdiction ended and the county's began. If we caught Keating's boys handling an accident in our territory, we'd waltz right in and take it away from them. They'd have their tow truck buddy on the scene, and we'd call our guy. There were times when we'd actually have two tow truck drivers hooked to the same car.

My main tow was a guy known as "Johnny the Hook." He had a big company with ten or fifteen trucks going all the time. When he and I first met, he didn't have a single towing contract, but I liked him and I started helping him out. I got him the contract for Willow Springs, and pretty soon we were hanging around together.

Aside from being in the towing business, Johnny was also a NASCAR champion driver. He was a real pro, with tremendous guts, the best I'd ever seen, and after a while we began racing together. Race car driving is a very expensive hobby. I knew from my own personal experience that Johnny needed a lot of dough to fund his activities at the track, so I helped him pick up towing contracts with different towns. He never kicked any money back to me. If he'd tried, I wouldn't have taken it—Johnny the Hook was my friend.

Thanks to all those contracts he'd picked up, Johnny was able to continue racing in a big way, plus, he and his family started living very well. Police towing is huge money, especially if you have a body shop and handle the wrecks. If you're also preferred by the insurance companies—which Johnny was—you're set. Every car he towed was his, from start to finish. The adjuster came right to his shop. When you do business like that, you can really rock and roll.

Before long, Johnny's company was doing the towing for

Bridgeview, Justice, Hickory Hills, Willow Springs, Hodgkins, Bedford Park, and Summit. He was making some big bucks on those towing contracts. Plus, he had car lots all over the area.

Johnny also became one of the major players in Chicago's chop shops. Basically, a chopper steals cars, takes them apart piece by piece, and then sells the dismantled parts—for way more money than the whole car would bring. Among the choppers, it's common knowledge that a lot of well-known race car drivers got into chopping as a way to finance their careers.

All chop shop owners have what's called *pullers*. Pullers are the guys who go out and steal the cars. Johnny had a bunch of pullers out on the streets. He was also buying wrecked cars and *tagging* them. Tagging is where you buy a totaled car for next to nothing and then go find (or tag) the identical car on the street and your puller goes and steals it. The chopper takes the incriminating— meaning traceable—parts off the new car and replaces them with the old junker ones. Presto, the chopper has a brand-new car on his lot without any possibility of being caught for auto theft. Back then, Johnny might sell a car for as much as twelve grand. Talk about a huge profit margin.

Sometimes I'd see a car Johnny was looking for, and I'd track the plate and find out where the guy lived. Then I'd call Johnny and fill him in. Later that night, one of his pullers would show up and haul the car back to the shop, where Johnny had a guy who could take a car apart in an hour and a half. That's saving all the glass, all the interior, everything. The car's gone, and all that's left is pieces. By sunrise, that stolen car didn't even exist.

In all the time Johnny was chopping cars, he took only one major arrest: he was charged with stealing about twenty-five vehicles, an offense that could have carried some very heavy time. I referred him to Alan Masters and they beat the case. In thirty minutes, Johnny was back in business.

When a guy chops a car, he ends up with a few parts he can't use. Either they're worthless or they're incriminating and junkyards can't take them. But the chopper still has to unload them, which isn't always easy. That's where I came into the picture. There were a ton of choppers in Cook County, and all of them knew I was a Willow Springs cop, with access to a canal, which just so happened to be a very convenient dumping ground. Helping those choppers out was one more way for me to make some extra bread. I'd seen Doc and Kresser do it, and they'd made a fortune. If it hadn't been toxic waste that contractor had been dumping they might've gotten away with it, too. Johnny and the other chop shop guys weren't dumping chemicals, so I figured what the hell—it seemed safe enough to me.

Some nights I'd sit by the canal for hours on end while those choppers dumped car parts. Primarily they dumped the cowl, which was where the serial number was imprinted. Cowls are as big as a dinner table, so they're pretty hard to hide. For the most part I had no idea what they were throwing in that canal. And I didn't really care. But eventually I would.

⊕ ⊕ ⊕

Jim Keating, the greedy bastard, had been watching me for a long time. Now I'm not saying I wasn't out for the money, too, but Keating was the type who would take what he wanted and then steal the rest. Keating knew the chop shop guys like Johnny the Hook were making millions. And once he realized I was in on the deal, naturally he wanted in on the action.

As luck would have it, Keating was in charge of the county's criminal investigation unit, the CIU. He ran the investigations into stolen cars, which led him to the choppers. They'd find some stolen parts and arrest a guy and refer him to Alan Masters. Keating and Masters worked together constantly. Masters would get the chop shop guy off with a fine and probation. After that brush with the law, the chopper knew he had to watch out for the cops, and that's when Keating and his boys made their move. They'd tell the guy he had to come up with a few grand every month from then on. They'd say they were watching him and that next time he wouldn't have Alan Masters to get him off the hook. Next time he'd be going to prison.

From what I heard, Keating's team clipped a chopper for anywhere from two to ten thousand a month, depending on what they figured the guy was doing. Keating had a whole operation going. They terrorized those chop shop owners. Unfortunately, it wasn't just Jim Keating and his pals the chop shops had to worry about. That was bad enough, but they also had the Outfit shaking them down. The two Jerrys—Scarpelli and Scalise—were on them real good. They were as cold as they come, so those owners didn't have a prayer. They're being shook down by the police. They're being shook down by the mob. They're paying their attorneys. It got to the point where some of them had to steal four or five cars a day just to take care of their expenses with law enforcement and organized crime.

It wasn't just doing time in prison the choppers had to worry about; guys who didn't pay up ended up dead. The extortion got to be unbelievable. A lot of choppers turned up missing; most of them were never found. Many of them were my friends, stock car drivers who raced on Saturday and Sunday nights and during the week chopped cars. Bob Pronger was a championship race car driver with an auto business in the Blue Island area. He was approached by Al Tocco's crew, meaning the two Jerrys, and told to pay up. But Pronger thought he'd be tough and refused to go along with the program, which works just fine until they put a gun to your head. Pronger was probably the first guy who got whacked as a result of the chop shop shakedowns. And talk about gruesome. They put him in acid and then disposed of the body out in Chicago Heights. Ten years later the cops found part of his skull and his teeth. That's all that was left.

Another chop shop murder caused such a big stink in the Outfit that it went all the way up to Tony Accardo. A relation of Accardo's wife had been whacked, and the family came crying to Accardo. But he couldn't have cared less; he washed his hands of the whole deal. The entire concept of chop shops and stealing cars was completely foreign to old-timers like Accardo. They had no idea of the money involved. But the young turks did. And after that—with the top guys like Accardo out of the picture—they just went wild. They didn't operate by coming in and trying to rough you up. They came in and gave you one chance.

Of all the Outfit crews into shaking down the chop shops, Tocco's was probably the craziest, and most brutal. Throughout the late seventies, they did some really off-the-wall stuff. In one instance, Scarpelli and Scalise actually drove a van right up to a guy's place of business. The doors slid open, and *boom*, they layed on that chopper, shooting him about nine times. That was the beginning of a reign of terror in Chicago. The newspapers called it the "Chop Shop Wars." When it was over, there were probably thirty guys dead, maybe more.

It was around this same time that I started hearing that Scarpelli and Scalise were planning to go after Johnny the Hook. By this time Johnny was one of the biggest choppers around, running a number of shops twenty-four hours a day and doing at least a hundred cars a week. So figure the money. Dough like that is just too tempting.

When Johnny heard that the word on the street was that he was next on the two Jerrys' hit list, he came to me. He was very concerned. He said, "Jesus Christ, Mike, what are we gonna do?"

I said, "What do you mean *we*?" I looked at him like he was nuts. "There's no *we* in this fuckin' deal."

But Johnny didn't hear a word I said. He knew I was pals with Scarpelli and Scalise—as well as a number of other Outfit guys. Because of that, Johnny always leaned on the idea that I'd protect him. He was totally one hundred percent confident of that. He couldn't accept the fact that there was only so much a guy in my position could do.

After a few weeks passed and Johnny still hadn't been approached by the two Jerrys, I was thinking that maybe they'd changed their minds. Then I found out they hadn't approached Johnny because he was my pal; they'd decided to approach me instead. When I met Scarpelli at the restaurant, I wasn't smiling. We both knew why we were there, so he didn't beat around the bush. He said, "Listen, your friend's got to pay something. Everybody else is payin'. He's been telling all the other owners that you're protecting him. The word's out that he's not payin' and it's hurting our business. It

My parents, Catherine and Joe Corbitt, made a handsome couple. *(Courtesy of Michael Corbitt)*

The Corbitt kids, looking like the all-American family of the fifties. From left to right: my big sister, Terry; me, all decked out in a cowboy shirt; my baby sister, Cecilia, who was the youngest at the time; and Regina. *(Courtesy of Michael Corbitt)*

At fourteen, I was thrilled when I received a shotgun as a Christmas present from Dad. *(Courtesy of Michael Corbitt)*

At my Sunoco I met a lot of mob guys, among
them Joe Ferriola *(left)* with an unidentified man.
Joe later became my friend and boss of the Outfit.
(Courtesy of Michael Corbitt)

Back in the early sixties, Turk Torello was
always hustling to make a buck for the
Outfit. *(Courtesy of Michael Corbitt)*

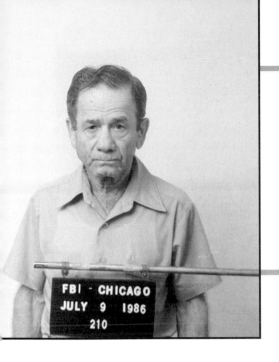

Turk and Joe introduced me to Vincent "The Saint" Inserro. Vinnie didn't look like much of a threat, but he was feared as one of the Outfit's most vicious enforcers. *(Courtesy of Michael Corbitt)*

Sam Giancana, my mentor and the boss of the Chicago Outfit from 1955 until his murder in 1975. When Sam got me a job as a cop in Willow Springs, Illinois, in 1965, he told me, "Don't forget who your friends are." I never did.
(© Chicago Sun-Times)

My father loved to see me in uniform and cherished this photo
taken of me, as a rookie, with my little brother, Tim. *(Courtesy
of Michael Corbitt)*

Doc Rust (here with his wife, Sophie) was boss during my tenure
as a young cop in Willow Springs, serving as the town's crazy
and colorful mayor from 1935 until his death in 1969. Doc,
who was no doctor, was probably the most corrupt man I've
ever known. *(Courtesy of Elmer Johnson)*

By the time I made chief of police in Willow Springs in 1973, I almost never wore a uniform—even when I posed for my adoring press after a high-profile raid like this one on an illegal-arms dealer. *(Courtesy of Michael Corbitt/photograph by Quentin Dodt)*

After making chief, I became Sal Bastone's constant companion. Sal was known by the FBI as a ruthless "money machine" who oversaw Chicago's multimillion-dollar vending and slot machine operations for Sam Giancana's associate and overseas partner, Hyman "Hy" Larner. *(Courtesy of Michael Corbitt)*

This photo of Hy Larner, taken during his testimony before the McClellan Committee in 1959, is the only known published picture of the Chicago Outfit's "mystery man." Larner would go on to become what many FBI insiders believe was one of the most powerful—and least recognized—individuals in organized crime during the twentieth century. (© AP/ Wide World Photos)

Meyer Lansky, an associate of Hyman Larner's and one of the most brilliant mobsters who ever lived. (© AP/Wide World Photos)

Carmen Bastone (Sal's brother) and his three sons caught on camera during FBI surveillance. Thanks to his association with Hy Larner, Carmen Bastone had it all: money, power, and, in his younger days, good looks. *(Courtesy of Michael Corbitt)*

Right: Victor Vita, Hy Larner's fair-haired boy in Panama during the seventies and eighties. A brilliant and handsome man, Chicago-born Vita carried a Panamanian diplomatic passport and spoke several languages. *(Courtesy of Michael Corbitt)*

Far right: Working vice in the Cook County Sheriff's Department was the perfect spot for a crooked cop like Jim Keating. *(Courtesy of Michael Corbitt)*

Charlie Nicosia *(from left with an unidentified man)* was a political fixer in the style of Outfit old-timers like Murray Humphreys. When Charlie asked for a favor, you didn't turn him down—something that put me on the hot seat with the mob in 1981. *(Courtesy of Michael Corbitt)*

Right: You didn't mess with Jerry Scalise, allegedly one of the Outfit's boldest enforcers as well as a brilliant master thief. *(Courtesy of Michael Corbitt)*

Far right: When it came to making a hit, Gerald Scarpelli was considered one of the best in the business. *(Courtesy of Michael Corbitt)*

William Hanhardt, Chicago's retired chief of detectives and mastermind of a nationwide burglary ring. *(© Chicago Tribune)*

Bill Hanhardt's ex-cop partner, Jack Hinchy. Hinchy was a close friend of mob wanna-be Chicago millionaire Joe Testa. *(© Chicago Sun-Times/Carmen Reporto)*

Chicago businessman Joe Testa *(far right)* and I *(center)* attended lots of social and political functions like this fund-raiser for Democrat Morgan Findley *(second from left)*. *(Courtesy of Denny Johnson)*

Right: Sam Giancana's notorious enforcer "Milwaukee Phil" Alderisio was among Joe Testa's Outfit pals and hidden "investors." *(© AP/World Wide Photos)*

Far right: Chicago hitman Marshall Caifano. In the sixties, Marshall targeted Joe Testa for extortion. Marshall didn't let up, terrorizing Joe for almost twenty years. *(Courtesy of Michael Corbitt)*

During my heyday as police chief, I made lots of VIP "pals"—among them, Chicago's mayor Richard Daley *(left)*. Unfortunately, Daley couldn't make the party/fund-raiser held in my honor shortly after my resignation. There were, however, several hundred other prominent citizens in attendance that night demonstrating their support. *(Courtesy of Michael Corbitt)*

I was a real cowboy during my eight years as chief of police in Willow Springs. I was always making the front-page news. But by 1980 the party was over; the village board wanted me out and I was fighting for my job. Finally, in 1982, I turned in my resignation. *(© Chicago Sun-Times/ Jack Lenahan)*

After leaving the force in Willow Springs, I became Sal Bastone's full-time sidekick. This photo was taken during one of our frequent Florida "vacations." *(Courtesy of Michael Corbitt)*

Right: Chicago's mobbed-up attorney Alan Masters bragged that he could get anybody off on any crime—if he was paid enough. *(Courtesy of Michael Corbitt)*

Far right: The disappearance and murder of Alan Masters's beautiful wife, Dianne, led to one of Chicago's most sensational trials. *(Courtesy of Michael Corbitt)*

This Cadillac, painted yellow with a white vinyl roof and vanity license plates, was identified as belonging to Dianne Masters immediately after being pulled out of the Chicago Sanitary and Ship Canal on December 11, 1982. Dianne Masters's decomposed body, discovered in the car's trunk, wouldn't be positively identified by authorities for several days. *(Courtesy of Michael Corbitt)*

My son Joey's christening in 1983 was one of the proudest moments of my life. Here, my sister-in-law, Gail Barone, cradles him in her arms while Joey's godfather, Sal Bastone, beams into the camera. *(Courtesy of Michael Corbitt)*

U.S. Attorney Tom Scorza was a relentless adversary of organized criminals, including me. *(Courtesy of Mark Alan Dial)*

The "Dianne Masters Murder Trial," as it came to be called in the press, was a real media circus. This courtroom sketch of *(from left to right)* Alan Masters, James Keating, and me appeared in the *Chicago Tribune.* (© *Chicago Tribune*)

Outfit business didn't grind to a halt when I was in prison; the FBI launched several investigations of Hy Larner and the Bastones during that time. In the nineties, FBI surveillance snapped this photo of *(from left to right)* Carmen Bastone and Hy Larner's personal accountant James Contis. *(Courtesy of Michael Corbitt)*

I didn't see much of my family while I was in prison. This photo, taken during a visit from my son, Joey, was one of my most treasured possessions. *(Courtesy of Michael Corbitt)*

One of the most respected men the Chicago Outfit has ever known, Tony Accardo served as boss and consiglieri from 1943 until his death of natural causes—a rarity in the mob—in 1992. *(Courtesy of Michael Corbitt)*

A greedy and cruel little man, Joey Aiuppa was Outfit boss from 1975 until he was sent to prison in 1986 for his role in the celebrated Strawman II case. *(Courtesy of Michael Corbitt)*

According to FBI insiders, Joey Lombardo is the current boss of the Chicago Outfit. Nicknamed "the Clown," there was nothing funny about Lombardo. Joey hated me from the start—mostly because I was a cop. *(Courtesy of Michael Corbitt)*

makes us look bad. I don't care what it is, a thousand a month, whatever, but Johnny's got to pay something. It's a matter of principle."

I was in a bad spot. I'd been doing business with Scarpelli and Scalise for years. I said I'd talk to Johnny, that I'd try to reason with him, but then I added, "You know what his answer's gonna be."

Scarpelli looked at me and shook his head. "Well, you gotta make him understand. He *has* to pay. We can't let this go."

I immediately went to Johnny and told him he had to pay those guys. But he was a hardhead. "I ain't payin' those jerk-offs nothing," he said. "If I see those two Jerry motherfuckers, I'll fuckin' kill 'em. If I see them around my family, I'll kill 'em." Johnny always carried a gun, so I knew he meant it—not that it would make any difference. I knew who'd win in that deal.

I went back to Scarpelli and gave him the bad news. He was real down about it, too. "You know what I gotta do now, Mike," he said. "I gotta go back and tell Tocco."

By this time, I was getting very uncomfortable with the situation; guys in the middle can get caught in the crossfire. I tried one last time to make Johnny understand what he was getting into. But it was no use. I knew they were going to whack him. I never came right out and said that, but I didn't have to. Johnny knew. Even then he didn't change his mind.

Several months went by and nothing happened, which was typical for those guys. Things were very tense for Johnny. One day I was driving in my squad car and I heard that there was a ten-alarm. It was at Johnny's shop. I drove a hundred miles an hour to get to that joint, and when I pulled up—well, I couldn't believe it. There was nothing. The whole block was gone. They'd pitched a bomb in the window of Johnny's chop shop and leveled it all. Stolen car parts were all over the place. It looked like a war zone. Fortunately, it was a Saturday, so none of the workers were there. But Johnny was always coming in on the weekends, so I figured that was it. The big bang.

I got on the phone and called his house. I was praying. When Johnny answered the phone, I wanted to cry I was so happy. I told him what had happened, and he drove right down. He took one look at that bomb site and he knew what he had to do. His family could've been in there. His entire crew of workers. Dozens of innocent people could've been killed. He said, "Get hold of those two assholes. Tell 'em I'll do whatever I have to do."

A few days later, Johnny met directly with Scarpelli and they got things worked out. Once he reopened his shop down the street at 79th and Harlem, he started paying Tocco's crew. And then, sure as hell, along came Keating's boys and they tried to extort him, too. But he refused to give them a dime, and they actually backed off. As amazing as that was, I knew there had to be a logical explanation, like maybe Keating had called a truce with Tocco's crew. Or maybe they'd cut a deal. I was beginning to suspect that there was more to the chop shop extortion game than met the eye. Like maybe Jim Keating was working with the mob.

⊕ ⊕ ⊕

There was one guy in the chop shops, Sammy Annerino, who was very stubborn, a real tough guy. They called him "The Mule." Everybody thought he was crazy. They said he was so bad he pissed fucking ice cubes. Even the top guys like Tony Accardo were afraid of him.

The Tocco crew tried to shake Annerino down, and he laughed in their faces. So naturally, they decided to whack him. The two Jerrys were on him big-time. They tried everything, but he was smart and very paranoid. They just couldn't get close enough to make a hit. Then the rumor came back to me that Tocco had come up with a solution: they'd team up with Keating's crew and go after Annerino together. If that wasn't what happened, then the circumstances around his death were a miracle of coincidence.

The day Annerino was murdered, a county guy on patrol called in the plate number for a silver Mercedes convertible, the car Annerino was driving at the time. Supposedly, the cop followed the car and led Scarpelli and Scalise right to Annerino. Fifteen minutes after the cop called in the plate number, the shooting started. From what I heard, they had to shoot Annerino about a dozen times with a shotgun before they finally killed him. They nailed him in the car, but he got out and ran down the sidewalk. So they got out and ran behind him, shooting him at point-blank range. Still Annerino didn't go down; he kept right on running and they kept right on shooting him. Finally he fell into a doorway, and they pumped about three more rounds into him. No wonder his name was "The Mule."

Naturally, the shooting was investigated. But nobody cared. Sammy Annerino was just another cockroach off the floor. Sal said, "Annerino was a fuckin' piece of garbage. The cocksucker needed to get killed. He's been askin' for it for years."

Whether the hit on Sammy Annerino had been a joint effort between Tocco's gang and Keating's county boys or not, it was getting pretty obvious that if one of their representatives paid you a visit, you'd better get with the program. Of course, some of Tocco's crew members, among them Butch Petrocelli, didn't want anything to do with the cops—especially if that meant they'd have to share the take with Keating's team.

Around the time Annerino was murdered, Petrocelli started getting a lot of complaints from his chop shop guys. It seemed they didn't mind paying him for protection, but having Keating's coppers shake them down as well was just too much. When they started telling him that they were tapped out, that they couldn't make their payment to him that week—because Keating's boys had been there first—things got serious. Butch Petrocelli wasn't about to stand still while somebody cut in on his action. He put the word on the street that he was putting a contract on Jim Keating.

It didn't take Keating a week to find out about that contract and who was behind it, and there was no way he was going to let a lowlife like Butch Petrocelli push him around. Supposedly he and his boys got in an unmarked car and went looking for Petrocelli. They pulled him over by a warehouse and dragged him inside, where they tied him up and blowtorched him for a while. Then when they got bored, they shot and stabbed him a few times. One of the guys—I don't know if it was Keating or not—strangled Petrocelli after that. It was total torture.

Butch Petrocelli's body was later found on Blue Island in the trunk of a stolen car. I heard afterward that it was one of his own pals, a fellow crew member, Gerry Scarpelli, who put the bug in Keating's ear. Having Butch off the street meant less competition for Scarpelli and Scalise. Plus, Scarpelli scored some major points with Jim Keating. They got very tight.

After Butch Petrocelli's murder, my suspicions about Jim Keating working with the Outfit grew. I was hearing more bad stuff about the guy. Keating was supposed to be a cop, not a stone killer. But that's what I was hearing: "Watch your back around Jim Keating." According to the other guys, with Keating, friendship meant absolutely nothing. He had a lot of chop shops pals through the years, and several of them ended up dead. I'm not saying Jim Keating did them all, but I do know about one of them for sure because Keating told me about it himself.

Timmy O'Brien was probably the biggest chop shop guy ever, with maybe thirty pullers on the street a day, easily making millions of dollars. He and Keating were like bosom buddies—until O'Brien started to think Keating was taking advantage of him. And the truth was, Keating had been taking from O'Brien with both hands for years. Around the same time his friendship with Keating was going south, O'Brien started getting some major heat from the feds. He was going to be indicted in what was pretty much an open-and-shut case. O'Brien was already very unhappy with Keating and the situation he was in, but now he'd had it. He told Keating he was getting out of the chop shop operation completely.

O'Brien had made his money. It made sense to him. It didn't to Keating, particularly when he got wind of the fact that O'Brien was going to turn and had a black book showing the payoffs he'd made to cops, lawyers, judges, and so forth. From what I understand, Keating called O'Brien and asked to meet him at a restaurant. O'Brien went—and here's a big surprise—a few days later they found him in the trunk of a car. Years later, Keating told me himself that he'd met O'Brien and that they'd ended up at a car lot. Keating said they talked things over real nice, that O'Brien never suspected a thing. After all, they were still friends. When O'Brien started to walk out the door, Keating came up behind him, pulled his coat over his head, and popped him.

I believe Jim Keating was telling the truth about the murder of Timmy O'Brien, but I have to say it was probably one of the only times in his life he was honest about anything.

⊕ ⊕ ⊕

"Chicago's Finest" have always had a love affair with organized crime. A lot of the cops I knew in the seventies were willing to turn their heads if the price was right. But being a team player is very different from being on the same team. Supposedly, Jim Keating wasn't the only cop who enjoyed that type of action. Word had it that Bill Hanhardt had an "arrangement" with Tony Spilotro—which didn't surprise me. Ever since my pal Roger Douglas's death, I figured he was the city's version of Keating and his Cook County crew. So how big a leap was it to be working with Tony Spilotro? When I mentioned it to Sal, he didn't say yes or no—which to me always meant a yes.

Without question, Tony Spilotro was a stone killer, but at heart he was really a burglar and thief. And as far as Bill Hanhardt was concerned, he'd been working both sides of the burglary racket for years, so he was made to order for any scheme Tony happened to dream up.

The Outfit had always been involved in stolen jewelry, with most of it being transported between Chicago and Las Vegas. Over the years I'd carried several packages of jewelry for Hy Larner, and I wasn't the only guy to do that by any means. Evidently it was this particular side enterprise that gave Tony Spilotro the idea to start his own jewelry theft ring. Unfortunately, he just forgot to let anybody in Chicago know about it.

In late 1976, Tony opened Gold Rush, Ltd., a jewelry store right off the Strip, and put "Fat Herbie" Blitzstein and his own brother, John, in charge of the business. What no one realized at the time was that Gold Rush was going to be a front for Tony's burglary ring. Once he had the store up and running, Tony started putting the operation together. He already had most of his old Chicago pals out west, so now he set them up in various locations. He sent Chris Petti to San Diego, Joey Hansen to Los Angeles, and Paulie Schiro to Scottsdale. In Las Vegas, Frank Cullotta and Sal Romano were his main guys.

Tony developed a real cozy relationship with the Clark County sheriff's Organized Crime Unit. Supposedly, three Las Vegas officers, including a detective named Joe Blasko, worked with Tony's crew. From what I was told later, Blasko in particular got very involved, acting as a lookout on heists and passing along the names of informants and FBI documents so that Tony could stay one step ahead of the feds.

Pretty soon Gold Rush was the number one joint in the country for fencing stolen jewelry. Tony had men in a half-dozen states—as many as fifty jewelry thieves—working in the operation. And believe it or not, nobody back home in Chicago had the faintest idea what he was up to. Of course, they should have known. After all, Las Vegas was a thief's dream come true, so it should have been obvious that overseeing the Outfit's interests out there wasn't going to be near enough to keep Tony occupied for long. And when a cowboy like Tony has too much time on his hands, it's a sure bet there's trouble ahead.

⊕ ⊕ ⊕

Ever since Sam's murder in 1975, when the papers slapped Tony Spilotro's picture on the front page and asked the public if this was the face of Chicago's next boss, Tony had been a hot property—and in more ways than one. With some of the old guys like Aiuppa, he was the fair-haired boy. With the FBI, he was Public Enemy Number One—they were on him big-time. So were the Nevada Gaming Control Board and the IRS. They all wanted a piece of Tony Spilotro. But every time they thought they had him, the guy skated. Forget John Gotti, Tony was the original Teflon man. He was also a potential loose cannon. Sal credited Tony's boss, Joey Lombardo—who was also Marshall Caifano's boss—with keeping him in line.

But I was hearing that he was getting harder and harder to control.

In the two years after Sam's death, there'd always been some problems to deal with, but by 1978 most of them were due to Tony's high profile; he was acting like the Don of Las Vegas. On one of my first trips out west that year, I heard through the grapevine that the Stardust was having problems with the Gaming Control Board. To make matters worse, Tony was always in the joint, throwing his weight around, pulling some stunt—which didn't make it any easier for Allen Glick and Lefty Rosenthal to run a tight ship. Keeping an operation together like the one the Outfit had in Las Vegas was hard enough without a guy like Tony making things more difficult.

By the end of 1978, Glick and Rosenthal would be forced to move on. Fortunately, it didn't take Chicago long to come up with their replacements: Al Sachs and Herb Tobman were submitted to the GCB as the new licensees for the Stardust. Sachs was an old pal of Les Kruse and very close to Hy Larner.

Al Sarno at the Dunes was also having his share of problems thanks to Tony. The Gaming Control Board was threatening to take away Sarno's license if he didn't throw Tony and his gift shop off the premises. From what I heard, Sarno would've done it, too, if it hadn't been for Joey Aiuppa coming to Tony's rescue. I think Aiuppa figured that throwing Tony out would just give everybody more negative publicity.

And that was bad enough as it was; Tony had been in the press almost constantly throughout the seventies. The papers were always having a field day with Tony: there'd been the murder trial of Leo Foreman, the brutal murder of Tony's codefendant in the case, Sam "Mad Dog" DeStefano, and on and on.

"They just can't afford to dump the guy," Sal told me one morning when I shoved the latest newspaper exposé on Tony Spilotro's notorious career across the restaurant table. "He's too fuckin' good at doing their dirty work." Sal gave me a look—the same one he'd used years before when he'd alluded to Tony's role in Sam's murder. The truth was, Tony Spilotro was a hero to some people; he'd eliminated what they'd considered a major problem: Sam Giancana.

Before Sam's murder, most of the Outfit guys—including Hy Larner—had blamed him for their problems, particularly in the gambling business, where the skim had been going down for several years.

If Sam had been responsible for the Outfit's declining finances, then you would've thought that four years later, everybody would've been shitting in tall cotton. But no, by the time 1979 came rolling around, they'd still be crying the blues. And the funny thing about it was that Hy Larner would be among the loudest.

CHAPTER 24

I t would be one of the world's biggest understatements to say that 1979 was not a very good year for Hy Larner. Or for just about anybody else in the Chicago Outfit. It should've been a great year for Hy, a banner year. Business in Central America had been going extremely well. In Panama, Hy was involved in several lucrative partnerships—resorts, casinos, cruise ships—with the country's key players: commander of the National Guard General Omar Torrijos and his loyal henchman and chief of military intelligence, Lieutenant Colonel Manuel Noriega. Guatemala was still chugging away, bringing in as much as twenty million a year in gambling profits. And Hy had started expanding operations to other countries south of the border: Belize, Nicaragua, Colombia.

In the Philippines the Outfit's longtime supporter, "mandated" president Ferdinand Marcos, was Hy's partner in several gambling rackets. There was the island of Macau, the gambling crown jewel of Southeast Asia. The Canary Islands. Spain. Australia. With Hy in charge, the Outfit was making considerable inroads in gambling rackets all over the world.

Of course, moving those gambling rackets into one country after another was an enormous undertaking, something I saw firsthand in Panama. From what I understand, the way Hy's gambling operations in Panama were organized was pretty typical of his enterprises elsewhere. Generally speaking, after a country was targeted, Carmen located a well-connected native attorney. Like Sam's man in Mexico, Jorge Castillo, this attorney would identify the appropriate parties—usually government officials—and smooth the way for the Outfit's entry into the country. In many respects this attorney was a glorified bagman, but he was also one of the most critical links in the organizational chain. For as long as the operation existed in his country, this man would serve as a buffer and troubleshooter whose main responsibility was shielding Hy Larner and his American associates from the glare of foreign publicity as well as any difficulties that might arise with the government. For that, he was paid very, very well—often being given a partnership in the operation in exchange for his loyal service.

The more I learned about Hy Larner's operations, the more amazed I was. When I found out that he was supplying U.S. military bases with slots—as many as a thousand machines—I could hardly believe it. I probably shouldn't have been surprised; supposedly a number of shady characters had supplied U.S. military mess halls and clubs with slot machines since the days of Prohibition. Back then, all the regular scams—skimming, rigging, and so on—were in full force. But in 1951 the Anti-Slot Machines Act was passed, putting an end to gambling on all military facilities within the United States. Slot machines remained on military bases overseas until the Vietnam War, when the Investigations Subcommittee of the Senate Committee on Armed Forces held hearings on machine corruption and fraud. As a result, the army and air force voluntarily removed all slots from their overseas bases in 1972. The navy and the marines, with over fifteen hundred machines on their overseas bases, continued operations.

According to Sal, the military had originally contracted with Bally to provide machines as well as their ongoing maintenance, but Bally didn't have the manpower at the time to send people all over the world to maintain a few machines, so they subcontracted it out to Hy Larner. That meant Hy had his foot in the door at all those bases. Although Bally continued to supply the traditional slot machines, once Hy's men got inside a military base, they started developing their own opportunities. Pretty soon Hy's companies were moving dozens of poker, bingo, and racetrack slot machines— all manufactured by Hy Larner in Miami, Chicago, or Kentucky— into bases around the world.

Thanks to the problems in Vietnam, there'd been a lot of heat about machines on bases, but Sal said things were beginning to look up. From what he told me, the marines and the navy were wide open; we'd been supplying them with slots for quite a while. And now, Sal said, despite their supposed voluntary ban on slots, the army and air force were starting to see the logic in working with Hy Larner—after he'd "greased a few palms in the Pentagon."

I understood from talking to Carmen that it could be a real challenge supplying so many overseas customers. Aside from the fact that the logistics were a nightmare, those machines were not cheap. One machine might go for ten grand or more. And we're talking thousands of machines. Carmen told me Hy had been trying to set up factories overseas, in cooperation with Bill O'Donnell at Bally Manufacturing. The guys at Bally, Carmen said, were helping set up a network of contractors in Spain, Japan, and the Canary Islands to supply machines and parts, but they weren't all up and running yet. By the close of 1978, we'd added a dozen military bases from all over the world to its list of customers. And there were dozens more to come.

Without question, the money derived from our relationship with the military was going to be tremendous. Still, Hy Larner wasn't one to rest on his laurels. He was targeting Costa Rica for his next big venture: a plush gambling resort on an island right off the Central American coastline. Of all the deals Hy had going, this one in Costa Rica had Sal the most pumped up. In fact, all through the Christmas season, that was all Sal could talk about. He told me that Hy said Costa Rica was going to be "as big as Panama someday." Like the island he and Noriega had there, Contadora. And even better, from Sal's perspective, because Hy had promised Sal that he'd be more involved in the Central American operations in the coming year. For a guy that hardly ever smiled, Sal Bastone was walking on air.

It was very exciting, like the sky was the limit. And it wasn't just Hy's gambling business overseas. The machine business in Chicago was going gangbusters, too. The money they were making was unbelievable. We'd put thirty fifty-five-gallon drums in an armored truck—all of them full of dimes, nickels, and quarters. That's a lot of cash.

The money came from machines all over Chicago, but it didn't come back to Zenith Vending. It never went through Zenith's books. It went to the Federal Reserve in Chicago and from there was wired to banks in Panama and the Grand Caymans. Some of it Hy passed on to guys like Ferdinand Marcos, Manuel Noriega, and the Shah of Iran. And, although no one ever mentioned any names—which was the way everybody did business in the Outfit—I heard that some of the money went straight to U.S. officials and politicians, as well as the military and intelligence guys who were so critical to us getting into some of those foreign places to begin with. From what I gathered, we're not talking slouches. These were all top guys.

Hy Larner didn't always pay guys off in cash. At Christmas he went all out with gifts, mostly jewelry. He sent couriers out with packages to friends all over the world. That particular Christmas we had a bunch of Panamanian military guys, mostly colonels, fly up to Chicago for a shopping trip. I believe we put them up at one of the Pritzkers' hotels—a Hyatt. It was common knowledge that most of the Pritzkers' financial backing at that time came from the Teamsters, meaning pension fund manager Allen Dorfman. By 2001 the Pritzker family's holdings would be worth over twelve billion a year and include not only the Hyatt chain of hotels but a share in Royal Caribbean Cruises, as well as a credit reporting company, Trans Union, and numerous other lucrative enterprises. Whenever possible, Hy wanted us to put his guests up at a Hyatt; he had this take-care-of-our-own policy.

To be on the receiving end of that policy was just tremendous;

Hy Larner knew how to repay a favor. In the case of the Panamanians I was nursemaiding that Christmas, he'd set up a half-million-dollar credit line at Rocca Jewelers for them. With that kind of dough at their disposal, those Panamanians had a ball. They came out of Rocca's just loaded down, with three or four briefcases full of watches, diamond rings, bracelets, whatever. All the finest, too.

Naturally, some of the rings they'd purchased had to be sized, which meant they would have to be delivered to Panama at a later date, after Hy's guests had gone home. That's how I ended up taking my first trip to Panama.

We didn't need visas or passports to go to Panama. No ID, nothing. Sal and I flew down to Miami, got on one of Hy's private jets, and went straight into Panama, where we landed on a military base. Forget about customs. We were treated like the King of Siam. Still, it wasn't much of a trip. I don't think we were there twenty-four hours before Sal had us on the plane headed back to the States. He wanted to get back to his family because it was the holidays. But he knew I was disappointed. When we got back to Chicago and hit that winter weather, he told me not to worry, I'd be heading back to Panama soon enough.

I figured Sal was probably right about that. After all, I was practically in the middle of everything those guys were doing by that time. Like the rest of Hy Larner's crew, I was flying high. I never felt so great.

⊕ ⊕ ⊕

On New Year's Eve, Sal dropped by the police station and gave me a bottle of Dom. Despite his weak attempt at putting on the holiday cheer, I knew something was wrong. I could tell just by the look on his face. Sal wasn't the same happy guy he'd been the day before, when he'd filled me in on what was happening in Costa Rica. The old serious-as-death-and-taxes Sal was back. I gave him a look. But he just shook his head and headed straight for the door. "Don't even ask," he said and walked out.

Sal put a real damper on my New Year's Eve that year. I didn't go out, and I didn't bother to open the champagne. My wheels were turning all night. I was going nuts wondering what the hell was going on. I could sense that this wasn't the same situation as when Sam got hit. But I figured it had to be business-related—which I thought could be just about as bad, probably worse, for me personally. I wondered if maybe it was Tony Spilotro. Yeah, maybe "the little guy" was in trouble out in Vegas again.

But logic told me that wasn't it. Things had been going fairly well in Vegas lately, all things considered. I'd traveled there for Sal at least five times over the past year and I hadn't gotten wind of anything of a serious nature. True, there'd been the problems with the Gaming Commission. And true, Chicago's revenues were said to be down. But the Outfit was still skimming a couple of million bucks a month out of their casinos, which was enough to keep the old guys from squawking.

Actually, I'd thought things were smoothing out—thanks to Hy Larner and what I perceived as a turnaround in the business in Las Vegas. In 1977 the Outfit had turned over all of Chicago's gambling interests, including Las Vegas, to Hy. More than anything else I think that demonstrates just how powerful he really was. At that time, just two men controlled organized crime in the United States. One of them was Hy Larner and the other was Meyer Lansky. Forget about the Italians. The Italians had fifty years to get it straight and they'd turned out to be a bunch of goons and leg-breakers. But Hy Larner and Meyer Lansky were geniuses. They were also real sly—which was why, shortly before New Jersey voters legalized gambling in November 1976, they'd gotten together down in South America to talk things over. Sal told me later that both Larner and Lansky had concerns about how things would shake out when Atlantic City opened up.

Everyone had managed to share the wealth in Las Vegas for a number of years without running into any major problems—although there wasn't any doubt that Chicago had pretty much edged out the competition. Even so, there hadn't been any serious conflicts. With Atlantic City happening, Lansky and Larner wanted to make sure it stayed that way. Mob conflicts meant shootings and murders, and that meant publicity. Publicity always hurt business.

As much as those two guys wanted to see gambling legalized in New Jersey, they also knew that Atlantic City could throw everything out of whack. "It would've turned out like something out of the Wild West," Hy Larner told me later. "Like a damned gold rush. Every one of those guys, from every city and every Family, would've been out there on the Jersey shore waving a loaded gun. We couldn't let that happen."

So Hy and Meyer Lansky came up with a plan: the Chicago Outfit would get Las Vegas, and everybody else, meaning New York's Five Families, would get Atlantic City. It doesn't take a Harvard law degree to see that this really meant that Hy Larner got Las Vegas and Meyer Lansky got Atlantic City. After all, those two men were the brains behind the skim at all the mob's casinos. The truth is, the man who handles the skim controls the money. He might let everyone think he's just standing on the sidelines, but he's the real power behind the throne.

Of course, when all the other bosses got together a few weeks later in Fort Lee, New Jersey, it wasn't positioned like Lansky and Larner were going to be in charge. From what I understand, Hy wasn't even there. I'm not sure if Meyer Lansky was, either. But that didn't matter, because they'd laid all the groundwork beforehand. Nobody squawked. Everybody agreed with their plan and that was that.

Ever since that meeting, Hy had been watching Tony Spilotro. I say "watching" because he hadn't given the slightest indication he wanted Tony removed. Not that I'd heard about, anyway—which was a strange thing to me. Sal had told me more than once that Hy wasn't satisfied with the way things were going out there. The only explanation I could think of was that maybe it was like Sal had said, maybe it boiled down to the fact that Tony was a very good enforcer. There was also the fact that even though the skim might be down, it wasn't like anybody was going broke. The guys in Chicago were still rolling in it, so their motivation to change things wasn't very strong. "You gotta be hurtin' before you'll risk rockin' the boat," Sal said. So to my way of thinking, that left out Las Vegas as the reason for Sal's gloom-and-doom attitude. I spent my New

Year's Eve wondering what the hell was going on. It wasn't until the following week that Sal finally filled me in.

"Things are turning to shit in Iran," he said as he got in the car. "Hy thinks the Shah may have to leave the country before it's over." Sal didn't need to say anything more; if the Shah was out, so were we. The casino Sam and Hy had put together in Tehran—the casino that Hy had turned into a six-million-dollar-a-year skimmer's gold mine since Sam's murder—would go down the toilet.

Like most Americans, I'd known there were problems in Iran. Still, I'd thought nothing much would come of it; I figured the United States wouldn't let its buddy the Shah be run out of the country like some common criminal. But on January 16, 1979, that's just what happened. The Shah took his family and ran for his life, going into exile in Morocco. Hy was not a happy man when he got that piece of news. But three days later, when the Ayatollah Khomeini, an Islamic fundamentalist, announced from Paris that he was returning to Iran to form a new government—well, then Hy went nuts.

As I was starting to learn, thanks to almost constant media coverage of the situation in Iran, things there had been coming to a head for some time. There'd been demonstrations against what many called the Shah's "repressive regime" for months. The previous September, during one of those demonstrations, the Shah's army fired on the crowd and hundreds of people died. But despite the increasing unrest, Hy and his intelligence pals hadn't been overly concerned. They'd figured the Shah's secret police, the Savak, would take care of any troublemakers. Evidently the CIA-trained death squad had been "taking care" of the Shah's detractors for years. And I figured that maybe that was the problem, that maybe the Iranian people just got tired of all that brutality and bullshit. Not that I told Sal that.

"It's like Havana all over again," Sal said after we got word that the Ayatollah had taken over the country's rulership. "Just like Batista and Castro. But this time the United States won't stand for it. We'll just have to wait until the Shah gets back in power."

Waiting was not one of Sal Bastone's strong points. He wasn't your most patient guy. Plus, he had to deal with Hy, who was now back in Panama and, from what I could tell, was going crazy over the Iranian situation. Aside from being a business partner, the Shah of Iran was Hy's friend. They were very close. So when the Shah started looking for political asylum, and President Carter didn't throw down the welcome mat, Hy was absolutely furious. But there was nothing he could do. His hands were tied. President Carter was not one of Hy's "boys." In fact, the way Sal told it, Jimmy Carter may have been one of the few guys in the world who wasn't.

According to Sal, with Carter in the White House, everything was going haywire, and not just in Iran. Carter's foreign policies were hurting the Outfit's business all over the world. The new president was implementing a complete reversal of the previous Republican administration's way of doing business with other nations, and thanks to that, there was a real shakeup going on down in Panama. During Republican George Bush's directorship at the CIA, Manuel Noriega had been on the CIA's payroll, making six figures. But the minute Jimmy Carter got into office, not only was George Bush out of a job at the CIA but so was Noriega. It would be another four years until the Reagan-Bush administration came into power and Noriega was back on the pad. During the first months of 1979, whenever Hy came to Chicago, we went over to the House of Hunan for lunch and listened to him rant and rave about President Carter's foreign policy. It got so I'd rather take a beating than sit through one of those lunches.

For months Hy was in a very sour mood. "The Shah of Iran has been a loyal friend of this country," he said. "And look how he's repaid . . . like he was a fuckin' communist, that's how." Almost worse, from Hy's point of view, was that President Carter had signed a treaty to return the Panama Canal to the Panamanian government by the year 2000. Every time the subject of the canal came up, Hy got so upset we thought he would have a coronary.

"That son of a bitch Carter is either stupid or a commie," Hy said when Sal and I picked him up at O'Hare that May of 1979. "Either way he doesn't get it. All that lousy cocksucker talks about is this humanitarian bullshit. You don't become a world power being Mother Teresa. That fuckin' guy Carter would give the farm away. He's so goddamned naive. You can't deal with these people like they were equals. They're not. The Shah knew that. He kept things under control. Another few years and Jimmy Carter will have the whole goddamned world in chaos. Just look at the way he's handled the oil deal. Made everybody conserve, made the American people suffer over some bullshit environmental crap he and some fuckin' hippies dreamed up. Thank God his term is almost over. He won't get another shot in the White House after what he's done. It's all set up. They're going to crucify him."

Sal gave me a sideways glance and said, with a hopeful smile, "And then maybe we can get back to business."

I didn't agree with all of Hy Larner's politics—the guy loved Paul Harvey's radio show, for Christ's sake—but I have to admit that I was happy to hear things might improve. Of course, shortly after we dropped Hy off at his condo, Sal knocked the wind out of my sails.

"Forget about what Hy said," he told me. "We're not out of the fuckin' woods yet. We still got Panama to worry about. Our deal in Iran may be over for all we know. Panama's our bread and butter. Supposedly, Torrijos has gotten a little too big for his britches lately. Hy thinks he's starting to believe all that democracy shit Carter hands out. Hell, for all we know, Carter could go in and fuck up everything with Noriega and Torrijos and we'd be out of business tomorrow.

"Those guys down there in Panama have got us by the gonads, Mike. Hy's been good to them, but so what? They'd fuck us in a heartbeat. Hy knows that, but he's too twisted up over this Iranian crap to think about Panama. He figures maybe we can ride this thing out until we get a new president in the White House. But in the meantime, what about Chicago and Vegas? What about our machine business? We can't be crying about the Shah of Iran while everything else goes to hell. We need to take care of business. We gotta look after Panama.

"So what if we lose Tehran? So we just go set up some new operations in some other countries. Shit, let's go after that resort deal in

Costa Rica. Let's lock it up. What the hell, there's a lot at stake here."

I hadn't seen Sal that fired up in a long time. And I thought he was right, too. There was a lot at stake. Hy Larner had built a gambling empire worth billions. And although we didn't know it yet—soon he'd be on the verge of losing it all.

CHAPTER 25

Sal and Carmen headed for Costa Rica in July 1979. Hy had paid these three top officials something like half a million bucks and promised them a percentage of the take when things got rolling.

Everything was ready to go. Just like in all the other countries where Hy had gambling operations and casinos, he had a front company set up that would be buying the machines and parts from Bally Manufacturing as well as from his own factories in Miami and Kentucky. At that time the front company for gambling operations in Central America and the Caribbean was Balicar, which was headquartered in the Grand Caymans. Juliano International was Hy's front company for operations in Panama and South America.

Once they got to Costa Rica, all Carmen and Sal had to do was be at the port when their shipment of three hundred slot machines came in by boat from Bally. Sal was so excited that he'd even been reading travel guides on Costa Rica, so he'd know his way around once they got there. He figured Hy was finally making good on his promise to get him more involved.

Evidently the trip down was pretty uneventful. They flew on one of Hy's private jets without any problems. They got a nice hotel. They got in their room, had a couple of nice meals, and the next afternoon they went down to the port to take possession of the machines. Hy had the wheels all greased, so getting them in through customs went smoothly.

By early evening, Sal and Carmen had the machines off the boat and in a warehouse. Everything was going like clockwork. Sal said he'd already started thinking about doing a little sightseeing the next day when, out of nowhere, this big gun battle started, right there on the street. Before they knew what happened, that warehouse full of slot machines had been seized by a bunch of crazy guntoting soldiers.

Political science wasn't high on Sal and Carmen's reading list; they had no idea who or what this gun battle was about. They'd later learn that Sandinistan rebels had set up jungle camps along the northern border of Costa Rica, with plans of attacking Nicaragua and removing its dictator, President Anastasio Somoza Debayle, from power. But at the time, Sal and Carmen didn't know that. All they knew was that these guys had the warehouse completely surrounded.

From a safe distance Sal and Carmen watched as the soldiers set up guards with machine guns to protect the perimeter. Needless to say, things were looking pretty bad. All the way back to their hotel they were hearing gunfire in the distance. That night they got word from the soldiers that if they wanted the machines back they'd have to pay for them, a ransom of fifty grand, U.S.

I can only imagine having to make that call to Hy. He had a very explosive temper. Just terrible. You didn't see it real often, and believe me, you didn't want to. Sal and Carmen were lucky they weren't with Hy when he got the bad news. I understand from talking to Victor Vitta, who was in Panama at the time, that when Hy heard about those machines he went crazy. But fifty grand in ransom money was chump change compared to what Hy had invested in those machines. They were worth more like a million, so paying fifty grand to get them back was nothing. So of course, Hy wired them the money.

Once Sal and Carmen got the cash together, things turned real cloak and dagger. They had to go to the warehouse at night to make the drop. And the whole way they were thinking that those sol-diers—who they now had identified as Sandinistan rebels—were probably going to take that dough and then kill them and keep the machines. Sal told me later that he really figured this was it for them.

But Hy Larner had already thought of that. He was always way ahead of those two guys. Just to play it safe—in case the rebels did try to pull a fast one—Hy had gotten in touch with the Costa Rican officials he'd been paying off for the last twelve months and told them it was time they earned their keep. He said he wanted Sal and Carmen—and those fucking machines—out of that country, pronto.

But Sal and Carmen weren't aware of Hy's plan to save them, so by the time they got to the warehouse, they were practically shitting in their pants. Carmen handed the money over to the rebels and held his breath. "It was the moment of truth," Sal said. "And it was actually lookin' like those sons of bitches were gonna play fair and give us those machines, too, when all of a sudden this fuckin' Costa Rican police patrol comes swoopin' in from all directions and starts shootin' up the fuckin' place."

So now there were two groups shooting at each other, and Sal and Carmen were right in the middle of it. But they weren't for long, because the police ran up and grabbed them, threw them in the back of a car, and away they went. They were terrified. "We didn't know what the fuck was happening," Sal said later. "We had no idea this was Hy's rescue mission."

They rode all night long, wondering what the hell was going to happen next. The next morning they got to a fishing village, where the police let them out and then drove away. They didn't have a clue about what they should do next. They were lost in a foreign coun-try, and even worse, they'd left behind a briefcase full of money and a warehouse full of slot machines.

They wandered around in that village until they found some-body who'd rent them a place. It was a little shack on the beach, and Sal said they were sitting in their room, discussing what to do next—my guess is they were screaming at each other—when *boom, boom, boom*—they heard gunfire again. And what do you know, the rebels were also in this village.

It was too dangerous to leave that shack. So they were trapped. As were dozens of other American citizens who had made it to that village—although Sal and Carmen didn't know that part. They didn't see anyone for days. They were holed up in that room, waiting it out. They were miserable, too, with no sleep and no food. One thing they did have, however, was plenty of bad water, which made them both sicker than dogs. Finally they got word that all the Americans were supposed to go down the beach to a certain spot that night where they'd be picked up by a U.S. military helicopter.

At that point, Sal said he and Carmen figured the worst was over. Knowing there were other Americans in the village made things seem less dangerous, like there was safety in numbers—an idea that fell apart the minute they started for the beach.

They'd never really considered how they'd make it out of the village with a war going on. And once they were in the street, it was obvious that all the Americans couldn't just waltz down the street together with bullets whizzing overhead. It was every man for himself, more or less.

It wasn't easy. Sal said they almost got killed half a dozen times that night, but finally they made it to the beach with all the other Americans. Helicopters flew in and transported everyone to a waiting U.S. aircraft carrier, and from there, Carmen and Sal hopped a military plane back to Panama.

As I later discovered, all the credit for Sal and Carmen's rescue couldn't go to Hy Larner. Hy had orchestrated their transport to the coastal village, but after that, Carmen and Sal had the U.S. military to thank for their rescue. As far as the military were concerned, the Bastone brothers were just two more American tourists facing danger in a foreign country. No one ever realized they'd saved two of the most important men in organized crime.

⊕ ⊕ ⊕

Sal and Carmen had been on a hell of a trip. And it would've made a terrific story, too, maybe even a funny one—except for one thing: Hy Larner had lost millions on that deal in Costa Rica. Forget about the cost of the machines—Hy had also lost a tremendous opportunity because of those Sandinistas. Just thirty days after Sal and Carmen's rescue, the Sandinistan guerrillas ousted the Nicaraguan dictator and—thanks in large part to their staging camps in Costa Rica and Honduras—managed to take control of Nicaragua.

There were other American investors who, like Hy Larner, were equally unhappy with the unstable situation in Central America and were willing to use any means to rout out the new Sandinistan government. By 1981 the United States under the Reagan-Bush administration would be secretly supplying arms to a rebel force comprised mostly of Somoza's old National Guard—the Contras. One aspect of the highly secretive military operation against the Sandinistas would later become known as the Iran-Contra Affair.

In the meantime, the situation in Central America was just getting worse all the time. With the exception of Panama and Belize, the entire region had become a hotbed of uprisings and revolts, putting an end to any hopes Hy ever had of investing in Costa Rica. That summer of 1979, he'd lost a deal that could have made him hundreds of millions. Believe me, nobody was telling that story at the dinner table. Hy Larner was not at all amused.

⊕ ⊕ ⊕

Hy hadn't been real amused that fall, either, when he learned how President Carter was treating Hy's old friend the Shah of Iran.

For most of 1979 the Shah had been a man without a country. He'd gone from place to place, trying to find somebody who would grant him political asylum. Hy was furious. Whenever he came into town he'd go into a tirade. "Treating a man like that is a disgrace," he said. "And after all he's done for the United States, too. Half the world would be communist today if it weren't for the Shah of Iran." Hy thought the Shah should've been considered a national hero because he'd been such a friend to the United States for all those years.

In October 1979, President Carter finally agreed to allow the Shah to come into the country. But only because the Shah had been diagnosed with cancer, and it looked like receiving treatment in the

United States was his only chance to beat it. It wasn't long before the whole thing turned into an international incident. In Tehran there were riots demanding that the Shah be extradited to Iran. Then they stormed the U.S. embassy and took a bunch of hostages.

When that happened, I expected the president to throw the Shah out on his ass. But Carter stood by his decision. The Shah stayed. If Hy was correct in anything he'd said about Carter, it was probably that Jimmy Carter really was a humanitarian. Of course, after the Shah's treatment was over, he was supposed to leave. And that was a problem: there wasn't a country in the world that would take him.

Meanwhile, in Tehran, things were going from bad to worse. Every night on the news it was another hostage story. They were saying American lives were in danger because the United States had been "harboring" the Shah. By this time President Carter was practically begging other countries to take the Shah. Still, nobody came to the rescue. And all the while things kept getting worse in Iran. The Ayatollah had the people over there all worked up. It was looking pretty bad for the hostages, and it was certainly a diplomatic nightmare for Carter. The whole time, Hy was livid. He was losing a fortune with that Tehran casino out of commission, and on top of it, his old pal the Shah was being treated like scum. Hy said that neither Nixon nor Ford would have stood still for what was happening in Iran, but Carter just didn't have the balls to do anything about it.

Finally I heard from Sal that things had been handled. He said a friend of Hy's—somebody "way up there"—had put in a call to President Carter, suggesting that the Shah could stay on Contadora, the Panamanian island where Hy and Noriega were developing some major resorts and casinos. Of course, Contadora wasn't actually in *their* names; like all the Outfit-controlled casinos in Las Vegas and every other deal those guys put together, they had a front man or two for Contadora. In this case I'm not sure who they were, but typically it's family members or friends who aren't directly connected to organized crime.

Evidently President Carter liked the idea of the Shah going down to Contadora because in December of 1979 the Shah went down to

Panama. I heard that Hy had his hands full. By the time the Shah got there, he was in pretty bad shape. Sal told me the guy was dying, and because of that, Hy wanted to make sure he had every luxury in his final days. Hy felt that was the least he could do. After all, thanks to the Shah letting them open up that casino in Tehran years ago, Hy was a millionaire many times over. Of course, so were a lot of other people. It was just that when the chips were down, Hy Larner was the only one who didn't run for cover. I was starting to see that sometimes that little Jew could be a real stand-up guy. And the way he tackled a challenge was really something, too. By the time 1979 was over, I'd actually admire the old bastard.

⊕ ⊕ ⊕

The challenges didn't stop coming at Hy Larner for the remainder of the year. That fall it was Las Vegas. And this time it was Tony Spilotro's turn to screw things up. Tony had already dodged a number of bullets, among them a major FBI investigation into his Las Vegas activities that had led to his arrest that past May. When the FBI's case against Tony surfaced and everybody found out what he'd really been up to, they were all pissed. It seemed that Tony had gotten more than a little out of hand; the FBI had tons of evidence to prove he'd turned Las Vegas into his own private betting kingdom. Aside from Gold Rush and his burglary network, Tony had set up a bunch of illegal sports betting joints all over Nevada. He was into loan-sharking and juice big-time. He'd actually set up his own crew to handle it all, even giving it a name: the Hole in the Wall Gang. The FBI had proof that Tony had been raking in a million bucks a month with that sports betting operation. And was "the little guy" sharing any of the wealth with his pals back in Chicago? Hell no.

The amazing thing to me, though, was that Chicago allowed him to stay in Las Vegas representing the Outfit. There were no repercussions. Those old guys were getting enough skim out of the casinos to keep them happy, and that was all they cared about. They didn't look down the road, so they weren't concerned. They figured, just like Sal did, that Tony would just hire the big-time lawyer in

Vegas, Oscar Goodman, and then "Oscar'll tear those FBI pricks a new asshole."

To this day, I can't get over it. Here was a guy who cheated Chicago out of all that dough from his sports betting deal, a guy who was under constant surveillance by droves of reporters and cops and federal agents, and still he managed to flourish in Vegas, not to mention stay alive. There was only one explanation: Tony was Joey Aiuppa's boy, his personal enforcer. With Aiuppa backing him up, Tony carried a tremendous amount of power.

Aside from having the Outfit's boss as his protector, Tony was simply one of the luckiest guys I've ever known. No matter what came at him, somehow he always skated. And that's exactly what happened with the sports betting charges. The FBI's case against him rested exclusively on records the Bureau had seized from his office. When the judge ruled in November that the FBI lacked probable cause for the seizure and ordered everything returned, Tony was off the hook.

Of course, a guy like Tony can't stay out of trouble for long; shortly after he'd dodged that bullet, Tony's name came up again. But this time there were serious ramifications for the entire Chicago Outfit as well as casino operations in Atlantic City. The FBI had evidence that Allen Glick, who was running a Las Vegas front for Chicago called Argent, had passed along almost two hundred grand to a Las Vegas cop, Gene Clark. Supposedly, Clark had taken up residence in Atlantic City and gotten a job as chief of security at a hotel owned by Penthouse International. Then he applied for a gaming license and made a sizable investment in a local casino, which meant the Chicago Outfit had just been caught with its hand in a very big cookie jar: Atlantic City. Retired officer Clark had accepted money from Allen Glick—an individual known to be an associate of Tony Spilotro and other organized crime figures—and then invested that money in an Atlantic City casino, a very serious no-no in the eyes of the New Jersey Casino Control Commission.

Although Clark denied any knowledge of Glick's ties to the Chicago Outfit or Tony Spilotro, the Casino Control Commission didn't buy it. Not only that, the commission was now concerned that other potential licensees, particularly Bally Manufacturing, might have ties to Chicago mobsters. In 1977, Bally had incorporated a New Jersey subsidiary, Bally Park Place, and started building a hotel-casino complex along the famous Atlantic City boardwalk.

Now, almost three years and three hundred million dollars later, Bally Park Place was set to open in December. All it needed was the Casino Control Commission's approval of its temporary gaming license. That was all that stood between Bally and a billion-dollar gold mine called Atlantic City.

The shit was about to hit the fan.

⊕ ⊕ ⊕

Sal Bastone didn't look happy. He had a piece of paper balled up in his fist and a major frown on his face. I watched him hurry across the icy parking lot toward the car. After he climbed in the car and slammed the door, he handed me a telex and said, "Get a load of this. Talk about some fuckin' bullshit. We are totally screwed. What the fuck are we gonna do now?"

The telex was from Bally Manufacturing's marketing director, Ross Scheer. It had been sent to Jorge Feldman, the president of Balicar in the Grand Caymans. Dated December 13, 1979, it was meant to serve as notification to Balicar that the question of the "true ownership of Balicar Limited and Juliano International, SA, has been raised and until the investigation is completed the compliance committee of Bally Manufacturing Corporation has advised me to instruct you that effective December 12, 1979, we will no longer honor orders—in house or not—for machines or parts."

By the time I'd finished the letter, I think my heart had stopped beating. Sal was right. We were screwed. All Hy's gambling operations throughout Europe and Central and South America had just been shut down. "Has Hy seen this?" I asked.

"Yeah. And he just about shit a fuckin' brick, too. He's already headed back from Panama."

"What about the Shah?"

"Fuck the Shah. We got a goddamned sinkin' ship over here. We gotta do somethin'. Hy has to intervene with somebody. O'Donnell, maybe. I mean, without Bally . . ."

I finished Sal's sentence for him. "We're out of business."

⊕ ⊕ ⊕

Bally Manufacturing was under the gun. It wasn't just Balicar and Juliano that Bally had cut off. Hy's other companies, Rentall and Flip Amusement, got the ax, too. Hy had front companies everywhere doing business with Bally. And every one of them got one of those telexes.

Thanks to the FBI's investigation of Tony Spilotro that previous May, the New Jersey Casino Control Commission had taken a real hard look at Bally's application for a gaming license. Things had already been tight in New Jersey, but now it looked like the commission was running scared. My guess is that the last thing they wanted was to be compared to the corrupt Nevada Gaming Control Board.

The most obvious problem for Bally was Hy Larner's lifetime friend William O'Donnell. Back in 1968, O'Donnell had changed Lion's name to Bally (taken from its famous pinball machines of the thirties) and gone public. Twenty years later, William O'Donnell was chairman of the board of a megacorp valued at over four hundred million dollars. And it would probably be worth twice that with the opening of Bally Park Place.

Overnight, everything hung in the balance. To receive its temporary gaming license in New Jersey, the commission informed Bally that its chairman, O'Donnell, would have to sever his ties with the company and put his stock in trust until the formal investigation was completed.

The gravity of the situation didn't escape Hy Larner. He went ballistic. He called for a meeting with Robert Harpling, director of slots for Bally Manufacturing in Chicago. With paranoia rampant in the company, Harpling refused to meet Hy at Bally's offices. Instead he wanted to meet at a joint called JoJo's near O'Hare.

All the way to the restaurant Hy was ranting and raving. When we got there, Carmen and Hy's Panamanian liaison, Victor Vitta, was already there waiting for us. We sat down, and about that time Harpling and Scheer walked in all dressed up in thousand-dollar suits. They hadn't sat down ten seconds before Hy had Scheer's telexes in his hand, waving them in their faces. He was screaming at them, too. *Screaming*. I was worried the owners would call the police.

"Who do you think you cocksuckers are?" Hy yelled. "Who do you think you're fuckin' with here? What do you think you're doing? You sons of bitches think you're going to put my fuckin' companies out of business? It's not my fault you're a bunch of dumb fuckin' pussies. If you think you're going to put me under just because you haven't figured who to pay off in New Jersey then you're dumber than I thought."

I think we all figured Hy was going to have a heart attack. Carmen was trying real hard to get him to calm down. Finally he ran out of steam and sat down, and they were able to get down to business. Over the course of the next several hours, Hy worked out a way to get around the problem. Since Bally couldn't supply us through the front door, they'd supply us through the back door. Hy would pay Harpling and Scheer cash for the equipment. Then we'd put our orders together, meet one of them at JoJo's, and pass it along. A few days later we'd send a truck to the Bally offices for the machines and parts. From there the truck would go either to O'Hare, where the material would be shipped overseas via commercial aircraft, or to the Palwaukee Airport outside Chicago, where everything would be loaded onto a private plane bound for Hy's Miami import-export front, Deverlandia, and flown out of the country from there. This was our MO for years.

So Hy Larner was back in business. Meanwhile, O'Donnell stepped aside as chairman at Bally, and according to Sal and Hy, the corporate officers told the New Jersey commission they'd ceased all business with us. Two years later, when Bally got its permanent license in Atlantic City, the heat was off. Sal told me that if a company got into problems working with Bally, Hy would just close that one down and start up another. What were any of those companies anyhow? Nothing but a fucking piece of paper. Sal said that was what lawyers were for.

And Hy had several lawyers at his disposal. One of them was his cousin Harold Lee. Another was Harry Slavis, who was a big mob lawyer from Chicago. Hy had them file a suit against Bally for shutting off the sale of equipment to Zenith. He wanted to make it look good, as if business had really come to a standstill. That way, if an investigation was launched, the feds would think they'd really put a crimp in things.

CHAPTER 26

Having a cozy relationship with the Outfit doesn't always work to your advantage. Take my friend Joe Testa for example. One look at him and you knew he had dough. You might have even thought he was connected. He wore suits that cost a thousand bucks. He had a limo and a driver and a gorgeous ninety-foot yacht on Lake Michigan. The people around him weren't just regular types, either—they were stars and big-name Vegas entertainers, the beautiful people.

Joe was a real estate genius who'd made millions and millions in property deals. He owned several fancy mobile home parks in the Midwest and some big resort developments in Florida. He also owned an S&L in Chicago, Sterling Savings and Loan.

I'd known Joe since 1969, around the time of Doc Rust's death. We'd been getting together every Wednesday night at Meo's for years. We'd have dinner, a few drinks at the bar. Meo's was known as a mob hangout, so there were always good-looking women in the joint. As Joe knew well, beautiful women were always attracted to a man with a pocket full of dough, but they went wild when they found out a guy was connected; being around a gangster really turned them on.

Of course, Joe Testa wasn't a gangster. He just wanted to be one. Generally speaking, Joe was a real gentleman and very polite. A shrimp of a guy, maybe five-three, he wasn't the type you'd take for a thug or a brawler. But if circumstances warranted it, Joe could be extremely tough, a real street fighter.

He had a terrible temper, like he was Rocky Graziano, or somebody out of *Goodfellas*. He'd go at a guy and just punch his lights out. It didn't matter where he was, either. When Joe got like that, it could be a restaurant, an airport, the White House, he didn't care. The other guy was going down.

One night on Rush Street we were in a washroom in this joint and I saw Joe just tear this guy up. All because the idiot was making fun of the way Joe talked. He was mocking him, like Joe was Don Corleone. Joe just lost it. He jerked that sawed-off little prick right out of the stall—he'd been sitting on the head. His pants were down to his ankles and his dick was hanging out. Joe slammed him against the wall—*thud*. Now blood started coming out the back of the kid's head. He slammed him again and then another *thud*. And now blood was going all over the place. Joe got him by the throat, he had him pressed against the wall. He got right up in his face and said, "So you think I talk funny?"

The kid didn't even get a word out. Joe put him down. It was just *boom, boom, boom*. One, two, three. And that kid was out. In under a minute Joe had him lying on the floor, blood pouring everywhere. He was crying, begging for his fucking life.

Joe said, "You still think I'm funny?. . . . Well, what about it, you fuckin' little jerk-off?" He gave that kid a real good kick in the ribs and then he turned and walked out. Believe it or not, he didn't have a single hair out of place. Not one drop of blood on his suit. Yeah, Joe looked fresh as a fucking daisy.

Joe Testa was tough all right. But tough that way and tough this way don't make you an Outfit guy. You've got to have a certain head. You don't just go after a guy when you're mad. You have to be able to kill him when you're laughing like a hyena at one of his jokes or right after your families opened a bunch of presents together under the Christmas tree. You've got to be that cold. And Joe wasn't. What Joe was, however, was a total mob groupie who would've done almost anything to get close to a made guy.

Joe's father had been good friends with Tony Accardo, a relationship that had given Joe his introduction to organized crime. Over the years, Joe had become friends with a lot of Outfit guys. They came over to his place for barbecues. He took them fishing on his boats. He went to their birthday parties, their anniversaries. He took trips with them, for Christ's sake.

Of course, Joe didn't pal around only with made guys. He had other friends, mostly cops, which on the surface might seem strange. But these weren't your normal police officers. They were guys like me, guys who liked to cut a few corners. That's what I figured anyway, until I happened to be at one of Joe's barbecues and ran into Bill Hanhardt, the rat cop out of Chicago, who according to my deceased friend Roger Douglas had been responsible for gunning him down. I wanted nothing to do with Bill Hanhardt. Unfortunately, Hanhardt's partner, Jack Hinchy, was Joe's close friend. "He's like a brother to me," Joe said when he introduced us. I never said a word, but after that I bowed out whenever I knew there'd be cops around. I was more comfortable with a group of hit men than with Bill Hanhardt, and I wasn't really comfortable with Jack Hinchy, either.

⊕ ⊕ ⊕

Jack Hinchy may have been like a brother to Joe Testa, but it was an Outfit guy—Felix "Milwaukee Phil" Alderisio—who was his idol. Before Philly (as Joe called him) went to prison, he managed to get Joe in way over his head with some of the Outfit guys' investments. Philly had done all sorts of financial deals with Joe.

Joe's relationship with the Chicago Outfit wasn't a one-way street. Thanks to Alderisio, the Outfit recognized the opportunity he presented, and they used him. With his savings and loan, they could borrow and deposit money without getting any heat from the IRS. I don't believe Joe ever said no to his pal Philly. And he didn't say no to any of his mobster friends, either. Pretty soon, it wasn't just one mob hit man taking advantage of Joe's S&L. Over the years, Joe invested millions of dollars for Outfit guys. He put up high-rises, low-rises—you name it. Most of it with their money, and most of it in his name.

Ask anybody and they'll tell you that doing business with Outfit guys is a very bad idea. These guys are the type who, like Sal used to say, "hate a fucking nickel because it ain't a dime." If the investment goes well, they'll start thinking you're screwing them, that nobody could resist skimming a little for themselves out of such a great deal. And if things go south, they'll think you stole it all. It's a no-win situation doing business with them. You may think you have a real cozy relationship, that you're actually friends—but if you get close enough to that flame, it's a sure bet you'll get burned.

⊕ ⊕ ⊕

In 1981 the Chicago Outfit was out of control. Tocco's crew had taken killing to a whole new level, so that whacking a guy didn't mean anything anymore. Forget finesse or discretion. Under cover of night or in broad daylight, it didn't matter. If they had a job to do, they did it. Guys were dropping like flies, the chop shop owners were still taking a beating, and the police departments were starting to look more like Outfit crews than crime-fighters. Politicians like Doc Rust's old pal Pat Marcy, from Chicago's First Ward, were operating more like godfathers than elected officials. And perhaps not coincidentally, cocaine was everywhere. A lot of the younger Outfit guys were dealing it—and doing it. They were living in the fast lane and dying there, too. Nightlife in Chicago meant disco bars, free sex, and fast highs. And if you were an Outfit guy, a fast buck.

I think it was Tony Spilotro who paved the way for a whole new ball game in organized crime. He changed the rules. He also brought down a lot of heat on everybody. So much so, in fact, that money was tight. Some of the guys were even getting desperate. If there was a debt—even the perception of one—you'd have to pay up. They were all relentless when it came to collecting.

One of the most relentless was Tony's predecessor in Las Vegas, Marshall Caifano. Marshall had made a name for himself as being every bit as ruthless as Tony. Back in the sixties when Marshall was in Vegas, he and Alderisio had been real tight. Philly was out there a lot and supposedly carried back a lot of cash for Marshall in those days, with the understanding it would be invested wisely—usually in some joint along their favorite nighttime haunts, Rush and Division Streets.

Nobody ever said Marshall Caifano was the smartest guy on the block, but he was ruthless. Since being exiled from Las Vegas, he'd served his share of time in prison. Thanks to his increasing age (in 1981, Marshall was sixty-nine) and the passing of the old guard like his pal Philly, Marshall had lost some of his stature. He'd also been under the gun financially.

For years, Marshall had made his nut by muscling big-money guys. He was very brazen. He and two thugs would march right into a major executive's office and, without any warning whatsoever, beat the ever-loving crap out of him. Marshall would threaten the man, his business, his family. And then, when he was laying on the floor, Marshall would tell him what he wanted. Usually it was big money, sometimes as much as half a million bucks. I heard Marshall had put the arm on five or six big-money executives. Once he got ahold of a guy like that, he just bled them dry. He'd be on them for years, constantly harassing them. Some guys actually went crazy.

Out in Vegas during the sixties, Marshall zeroed in on an Indiana oil guy named Ray Ryan. Ryan was a businessman who owned a Kenyan resort with actor William Holden. He was a big gambler, a real high roller, and Marshall figured him for an easy mark. Instead, when Marshall demanded sixty grand or else, Ryan just laughed in his face. When Marshall pushed harder, Ryan threatened to go to the FBI. A year later, thanks to Ryan, Marshall was convicted of extortion and sent up the river. But that didn't mean Ryan was off the hook. A few years after their confrontation, Ray Ryan turned the ignition in his car, and *boom*—end of story.

You would think this type of shakedown would eventually cause a problem for everybody. These weren't chop shop guys or burglars, they were legitimate businessmen Marshall was going after. You can muscle a guy who's breaking the law and get away with it—he's got nowhere to turn—but when he's legit, he has the option of going to the police or the FBI. And when law enforcement starts sniffing around, it puts pressure on the entire organization, slowing business. Nobody wants that.

I have no idea why those executives Marshall was shaking down didn't run to the FBI—maybe they were just too scared—but for whatever reason, the cops hadn't gotten involved. And as far as the Outfit was concerned, Lombardo and Aiuppa just looked the other way—probably because Marshall made sure they got a piece of the action.

Marshall was always looking for his next mark. He was very methodical in how he went after a guy, and very patient, too. He wasn't like Tocco's bunch, one chance and you were done. He liked to terrorize his victims. He'd work a guy for years. In fact, Marshall had been working my friend Joe Testa ever since Philly had died in prison in 1971.

Philly had given Joe Testa two million dollars to invest. After he died, Marshall went to Joe and asked him to give it back. Joe didn't feel the money belonged to Marshall, so no way was he giving it to him. He figured Philly gave it to him and that was that. Of course, Marshall saw it completely different. He and Philly had been partners in several deals, and Marshall believed that two million was part of the Vegas skim he'd sent back with Philly to invest. In any case, Marshall and Philly had invested in a ton of different things together—restaurants, porno, prostitution, gambling. So now, in Marshall's way of thinking, this was *their* money.

Up until then Joe had had no problem investing Outfit money. Guys like Tony Accardo and many others trusted him. An investor always got his money returned immediately if he asked for it. But none of this mattered to Marshall. He wanted his money.

At first Marshall toyed with Joe. He'd call and ask him for the money, and when Joe blew him off, he'd start trying to mess with Joe's head. He might mention Ray Ryan and try to put a little fear in him. Or he'd do stuff like tape a fake bomb on Joe's car—which, of course, Joe didn't know was fake until he called me and we had it checked out.

Joe Testa was way too tough to fall for simple scare tactics. Back in 1968, before Joe and I met, he'd shown all those Outfit guys just how tough he was, too. He had a club in Schiller Park, and at the time they were having a problem with American Vending, which was Louie Eboli's company, and the Bastones at Zenith. They both were insisting that Joe put their machines in his joint. Pretty soon they were fighting over several clubs in the area. Now the clubs started getting blown up. One night after closing, somebody pitched a bomb in the front door of Joe's club. It did a ton of damage, and Joe was enraged. He told both machine companies to fuck off and went out and bought his own machines. Shortly after that, Joe looked out one night and saw one of the vending enforcers parked in front of the club watching the joint. Joe went ballistic, grabbed a gun, and ran outside to put a bunch of bullet holes in that guy's brand new Cadillac.

They came back to him after that and he said, "I'm not gettin' anything from either one of you jerk-offs but a bunch of fuckin' grief. Don't ever come back around my fuckin' joint again or I'll machine-gun all of you motherfuckers." I think they knew Joe would have done it, too. After that, most guys were scared to death of him. They thought Joe Testa was nuttier than they were.

For seven or eight years after Philly's death, Joe was able to shrug off Marshall's threats. And why not? After all, they never amounted to anything. It was just Marshall circling, giving his prey a bump here, a nibble there. Then, sometime in late 1979, Marshall turned up the heat a notch. He dropped by to see Joe at his office in Sterling Estates. He looked around and dropped the hint again. After that, Joe started running into Marshall all the time. He might be walking out of a bar on Rush Street, and there was Marshall. Or he might be coming out of a restaurant with his wife, and what do you know, it's Marshall sitting in a car in the parking lot.

Then Joe happened to run into Marshall out on the putting green at the Brookwood Country Club. Joe told me later that by now he knew Marshall had become very serious about collecting the two million dollars. Joe was shaken up. He never played golf at Brookwood again, even though he owned a condo on the property and was a member of the club. I think what really bothered him was that he'd always thought he was part of the Outfit, that the shit with Marshall all those years was just some type of cat-and-mouse game. But now he saw that Marshall was serious. And it just really got him down.

But Joe still didn't pay up. He was too stubborn. It was "Fuck them. Nobody's gonna fuck with me. I'm not giving that lousy cocksucker a fuckin' quarter." Then one night Joe found another one of those fake bombs taped on his car. He was real pissed about it, too. He called me and told me to come down so I could see it for myself. When I got there, Joe was concerned there might be a real bomb somewhere on his car, so we left his car in the parking lot.

Later I sat him down and told him he needed to rethink his position with Marshall. I told him we both knew the guy was serious now. He'd give Joe a little more time—shit, it could be months—and if Joe didn't come up with that money, Marshall would come after him. I said, "Look, I've seen what these guys can do. Marshall's got all the muscle. You can talk all you want, but you got nothing, Joe. Not even a fuckin' crew."

But no matter what I said, Joe would not listen. And the truth was, Joe did owe *somebody* that money. It was Philly's money before he died. So maybe it wasn't Marshall's, but it certainly belonged to Philly's family or a relative. The bottom line was, it wasn't Joe's two million bucks. Still, he wouldn't budge.

By 1980 things had gotten very tense. Marshall's crew started blowing up stuff for real. They blew up Joe's restaurant. They blew up his trailer court. The bombings were in all the papers. The *Chicago Sun-Times* interviewed Joe about it and he said, "No bastard is ever going to muscle me. I'm not going to take it." Saying things like that to reporters is just asking for it. Marshall and his crew had something to prove after that. They couldn't let this jerk make a monkey out of them. And sure enough, right after Joe shot off his mouth to the press, Marshall came for him.

Joe was driving down the street one night, and he had his wife in the car. And here come these guys, they pull up by Joe, and *boom, boom, boom,* they open fire on Joe and his wife. They put around five deer slugs right through Joe's Lincoln. One of them went through the front seat and tore a huge hole in the upholstery. It barely missed his wife.

It was like something out of the old Capone days. Joe turned into a wild man. He slammed on the brakes and made his wife get down on the floor. Then he jumped out and started shooting back. Joe always carried a .38 snub. He got off six rounds before they took off—which was fortunate, because that's all the firepower he had.

That one incident got to Joe more than anything else, mostly because his wife was involved. He thought that was going too far. It wasn't long after that drive-by shooting incident that they blew up his office. And now Marshall came around again to pay Joe a visit. He got right in Joe's face. He told Joe it was him blowing up everything. He said, "I'm doin' it, Joe. I'm after you. And I'm gonna stay after you, too. We want our money." Joe just looked him in the eye and told him, "Fuck you. That money's mine now."

It was real stupid of Joe. He had two million dollars in the shoe box in his closet. He would've never missed that dough. They didn't even want any juice on Philly's money. But Joe told me that if he ever started paying Marshall, he'd be the guy's bitch for the rest of his life.

Joe was out of town when they blew up his house. It was a beautiful, expensive home in the suburbs, and it was totally destroyed. Fortunately, no one was there when it happened. But there could've been. Joe knew that. He called me and asked me to get it all arranged. He wanted a sit-down with Tony Accardo.

⊕ ⊕ ⊕

Don't get the impression that playing nursemaid to Joe Testa was the only thing I'd had to do those past few years. I was working with Sal's crew, running back and forth to Vegas, plus I was handling all the responsibilities that came with being chief of police in

Willow Springs. I also had a few bullets of my own I was trying to dodge.

Part of having friends in high places, in being connected, is the heat that goes along with those relationships. And talk about heat. There were so many corruption investigations into various aspects of my life that I couldn't keep up with them all. Charges I'd been selling badges, taking kickbacks, all sorts of stuff. And there was the IRS. They'd landed on me with both feet in 1975, after I built my house. They wanted to know where I got the money on a cop's salary for a half-million-dollar home with a swimming pool. I wasn't going to say I got it from Sam Giancana. Instead I told them I saved it in my mattress. They'd been on me ever since. Civil actions, criminal actions, you name it. The IRS had turned up stuff against me over the years, but they'd never turned up any hard evidence of wrongdoing on my part.

Looking back now, I can see that most of my problems I brought on myself. Not only in my extracurricular activities but also because I was so uncompromising. I drove the village board crazy for years. They were dying to get rid of me. They met twice a month, and every meeting was a war. They'd find certain things to bitch about and they'd attack me. It was always something. For example, they hated that I continued to race cars. They even went so far as to have a big meeting just to tell me I had to quit. They thought having the chief involved in racing set a bad example for the community's youth. I quit for a while, but then I decided, fuck them. I got a pro NASCAR stock car and went right back at it. I got all sorts of publicity, too. Which pissed them off even more.

I got a lot of heat over the way I looked, too. For one thing, I never wore a uniform. We're talking the late seventies, early eighties here, now what guy wants to walk around looking like a jerk? Not me. I wore a leather sports coat and open silk shirt with a bunch of gold chains, tight jeans, and cowboy boots. My hair was fairly long and I had it permed. I drove a Corvette. It was all about style. There was no way anybody would've picked me out as a copper. Which was just the way I wanted it.

Of course, the village board didn't see it that way. Every meeting somebody said that the citizens should be able to recognize me as the chief of police. They wanted me to wear a uniform. Finally I decided to have a little fun. I went and got a uniform made that would make you throw up: it was real bright blue with big fucking gold stripes right down the front of the pants. It had a matching Eisenhower jacket with these huge ugly gold buttons and gold trim. On the shoulders there were these big broom things, so it looked like something a Latin American dictator would wear. Naturally, I got a hat, too. It had all this scrambled egg shit and a big eagle on the band. The whole deal cost around fifteen hundred bucks. I could hardly wait to see their faces at that next board meeting.

The next month, I walked in with my new uniform. The minute they laid eyes on me, everybody in the audience almost went nuts they were laughing so hard. But I didn't crack a smile, playing it very straight. When it got time to come to order, the mayor couldn't conduct the meeting because nobody could stop laughing. It was hysterical, with the board silent and half of them unable to look at me. Finally somebody from the audience said, "We're happy to see that the chief is finally wearing his uniform." And the crowd just broke up. Of course, the board members didn't think it was a bit funny. Later I got the word never to wear that uniform in public again—which I didn't. Once was enough.

They tried to fire me several times, but it never worked. From the time I made chief, I'd always had the support of the mayor and the village attorney. But then, in April of 1981, all that changed. Willow Springs got a new mayor, Frank Militello.

Like every candidate for mayor since Doc Rust, Militello had promised he was going to clean up the town. The difference was, Militello actually did it. One of the first things he did when he got into office and met with the village board was fire me and appoint a new guy, Jim Ross, as chief of police. Fortunately, I'd gotten wind of the fact that this was coming down, so the very next day, me and my attorney, Dennis Berkson, went to Circuit Court Judge Joseph Wosik, who just so happened to be a close friend of Joe Testa's, and got him to issue a temporary restraining order against the village firing me. Now this Judge Wosik didn't care about who was right and who was wrong. We're talking about a judge who didn't wear pants under his robes to court, for Christ's sake. The fix was in and that was that. Right away he made his ruling and granted the temporary injunction.

I went back to the police station and immediately booted Jim Ross out of my office. Despite the headaches that went along with being chief of police, I knew I was far more valuable to Sal and Hy with a gun and a badge. I'd always worried that without those two things, they might consider me just another thug wanna-be. I was going to do whatever it took to stay in that office.

Of course, I knew the temporary restraining order would run out soon. I had to go back to court and get a ruling from the judge in order to be permanently reinstated as chief. But I was confident that the judge would rule in my favor. Which my attorney couldn't understand. But from my point of view, it was going to be a total lock, thanks to my friend Joe Testa and his relationship with the judge.

A few weeks later, when Joe asked if I'd arrange a sit-down with Tony Accardo, I couldn't turn him down. Sure, I hated being in the middle, but after all Joe had done for me, I figured I owed him at least that much—and maybe more. Just how much more, I was about to find out.

⊕ ⊕ ⊕

It was two o'clock in the morning. Everybody who needed to be there was there, sitting at this big table in the basement of the La Copanina: Tony Accardo, Joey Aiuppa, Jackie Cerone. And now Joe Testa started telling them his story. They didn't say a word. They listened to the whole thing.

Joe went back and forth between being real forceful and having big tears in his eyes. He told them, "Look, I can't take it no more. They're blowin' up everything I got. They almost killed my wife, for Christ's sake. Haven't I been your friend all these years? And what about all those trips to Florida, all the parties? What about all the shit at the savings and loan? Doesn't that mean somethin'? I've never asked anybody for a fuckin' thing. Nothin'. And now all I'm asking is that I get a little assistance in dealin' with this guy. Call him off. I can't take it anymore. You gotta help me."

When Joe was done, Accardo said, "I can speak for everyone here on what we're going to do regarding this situation. But first I want you to know, Joe, that you're my friend. Your father was my friend. We socialized together. We had a lot of good times over the years.

"But on this deal with Phil's money, I can't help you. You took the money to invest for him. We all know that. And then when he died and Marshall asked for it back, you refused. But you owe that money. It's not yours. So I'd say you're going to have to make up your mind. You have to do one of two things. You either lay them down or you pay them. Those are your choices. It's up to you. Either way, it's out of my hands."

In the car, Joe was very quiet. He suggested we go for some breakfast. He wanted to talk. He was starting to be a little paranoid by that time. He didn't want to talk in his house—he was living in his trailer court since the bombing. He didn't even like to talk in the car. We got inside the diner and sat down. We got some coffee. Finally he said, "I think I'm gonna take his advice."

I said, "Whose advice?"

"Accardo's. Yeah, I think I'm gonna whack those motherfuckers." Now I thought he'd gone over the edge. I told him, "Joe, are you crazy? Forget all that bullshit. You don't have a fuckin' crew. You gotta pay this fuckin' money."

He just shook his head. "No, I'm not gonna to pay. I'm gonna whack 'em, that's what. All of 'em. Yeah, that's what I'm gonna do."

At this point, Joe started telling me about how we could whack Marshall and his crew and then bury them on his farm in a back pasture. He said he could get a hole dug and the whole bit. And I could see he was serious; he was already plotting and planning. I told him, "I'm out of this deal. You hear me? And whatever you fuckin' do, don't talk to anybody about it."

He smiled and picked up the menu. "Do you think I'm fuckin' crazy? It's you and me. We're gonna do this. It's just you and me, buddy, on this deal."

I got up from the table and told Joe we were leaving. I took him straight home. He didn't mention his plan anymore that night.

Over the next few days I thought about what he'd said about taking those guys out. I knew Accardo had really disappointed Joe. He'd thought of Accardo like an uncle. But he understood what Accardo had meant when he'd said it was "lay them or pay them." Whatever Joe decided, Accardo was giving him his word that the Outfit would stay on the sidelines. Joe could pay up or he could whack Marshall and his leg breakers. There would be no repercussions from Accardo and his pals. It would be over. A sort of winner-takes-all, survival-of-the-fittest deal.

The more I thought about it, the more sense it made. Maybe by that time I had no conscience. Maybe it was my feelings for Joe. I couldn't just stand by and watch what he was going through and not lift a finger. I knew it was over if he didn't pay. I knew Marshall would kill him, sooner or later.

⊕ ⊕ ⊕

"I got it all set up," Joe said. "I told Marshall I got the dough. He's supposed to meet me this afternoon at the warehouse by my office. He's planning to pick the money up then . . . him and his two goons."

By this time I'd decided to join Joe on the deal, one hundred percent. It seemed the only way Joe was ever going to get out of this problem was to whack those bastards and get it over with. I was a good shot, so I knew it would be easy to take them out. Joe had the body bags at the warehouse, and there was a hole already dug out at his farm in Wisconsin. I'd gotten sandbags and lye. We figured we'd dispose of their car in the canal at Willow Springs.

We were ready to rock and roll. An hour before Marshall was supposed to show, I arrived at Joe's office. I walked in and—what do you know—Joe was lying on the floor. He'd had a heart attack. I got him in my car and rushed him to the hospital.

Talk about fate intervening. That heart attack not only put an end to Joe's plans to get rid of Marshall, it also signaled the end of an era. Once he recovered, Joe was a different man. Overnight, all the fight had gone out of him. When the doctors told him he had a choice—have open-heart surgery or die—he was devastated. You could see it in his eyes. You could see it in his hands, the way he'd shake. Joe had always been such a tough guy, it was hard to watch him go down like that.

It probably wasn't a week after Joe's heart attack that Sal started to harp on me about Philly's money. He said, "You know, your buddy Joe, he owes those guys. He might be sick, but he still has to pay up. You'd better talk to him."

Marshall didn't know that Joe's heart attack had saved his life; instead he figured Joe was using it to avoid paying him. Evidently Marshall got concerned that Joe was starting to make jerk-offs out of all of them, which was when the money became less important to Marshall than maintaining his stature in the Outfit. And when that happens with those guys, you'd better start looking over your shoulder, because somebody is going down.

By June 1981 I was beginning to think I'd been wrong, that maybe Marshall was going to back off. Joe had gone down to Florida to "get away from it all." He'd been bugging me to come down, maybe play a few rounds of golf, go out on his boat—just like old times—and with the judge's decision on my case due at the end of the month, I figured I'd take him up on it.

I don't think we slept from the minute I got off the plane. We had a great time. We hit the town every night, and during the day we found some girls and went out on the boat. We got half a mile out and it was everybody get naked—that was Joe. Sun and fun. Work hard, play hard. It wasn't quite the same as it used to be, but Joe gave it his best shot.

⊕ ⊕ ⊕

According to the police report, it was around noon on June 27, 1981, when Joe put the key in the ignition of his blue Lincoln Towncar. The explosion blew the windows out and blasted the hubcaps right off the car. It also blew off Joe's leg and hand. The force was so tremendous that it sent shards of metal, like shrapnel, in all directions. Panicked, Joe's pal Jimmy Aquilla struggled to get out of the car before the gas tank exploded; he made it just as the next explosion roared through the parking lot and consumed the car in a ball of fire. Hearing Jimmy's screams for help, some golfers rushed over and helped pull Joe from the flames.

When I got the news that Saturday afternoon, I flew down to Florida immediately. But I didn't know what to say once I got there. There was nothing to say. Joe was screwed.

I got on a plane to head back to Chicago knowing Joe was in very bad shape. Terrible shape. In my heart I knew he wasn't going to make it. He was weighing heavily on my mind when I walked into the courtroom on Monday morning.

Over the past weeks, I'd enjoyed every second of the hearing. Judge Wosik had really been on the village trustees. They'd put on witnesses who had made ridiculous accusations about my behavior, and Judge Wosik had told them to back it up, to show him some evidence. But they couldn't. Every single witness's testimony was just hearsay and gossip. They didn't have anything. They didn't prepare properly.

But my attorney had done a great job. We had terrific witnesses and all our ducks in a row through the entire hearing. So I should've been overjoyed. I should've been smirking at the very least. There should've been satisfaction at sticking it to those bastards on the village board. But now all I could think about was Joe and that bomb. Joe. Without a leg. Without a hand.

When it came time for the judge to make a ruling in my case, I didn't even care. But before he could say anything, Judge Wosik's secretary came in and whispered something, and he got up and walked out of the courtroom. He didn't dismiss the court. He didn't call for a recess. The judge's behavior was highly irregular and it sent everybody into an uproar.

In a few minutes he came back in and pounded the gavel. He was like a different person, with a completely different expression on his face—very serious and uptight. I started to wonder if maybe this wasn't a lock after all. Maybe something went wrong. I held my breath while the judge gave his decision reinstating me as chief of police in Willow Springs. And then I heard him say, real softlike, so you could hardly hear him, "This one's for you, Joe."

As I walked out of the courtroom I heard the news: Joe Testa was dead. Still, he'd put the fix in for me just like he'd said he would. I only wished I could've done the same for him.

⊕ ⊕ ⊕

I tried to piece together what I knew. It wasn't much. Jack Hinchy, Allen Dorfman, and Dick Gazie were down in Florida to see Joe and play a little golf. Jack Hinchy was a real golf nut, so he and Joe were always out on the course. But on the morning of the car bombing, Jack told Joe and the other guys to go ahead and hit the links without him. To my knowledge that was probably the first time Jack Hinchy ever turned down a game of golf with Joe Testa.

With Jack staying behind, Joe invited Jimmy Aquilla to come along and fill out their foursome. They hadn't been out very long when Jimmy, who wasn't a great golfer, wanted to call it a day. Dorfman and Gazie said they still wanted to hit a few more balls, so Joe and Jimmy said good-bye and headed for the car.

Dozens and dozens of people were interviewed in the case. As far as I know, Dorfman and Gazie were not. In fact, the papers didn't even mention Allen Dorfman's name. He was just "another member of the foursome." Everybody else was interviewed by the police and ATF. But not them. It was all very strange and, I thought, very suspicious.

⊕ ⊕ ⊕

Just before Joe's passing, another important person in my life died: my father. It was a hell of a loss. Dad had always been my biggest supporter. He'd been so proud of me being chief of police. I gave Dad a real good send-off. We had a big procession—a mounted patrol and a bunch of squad cars—the whole nine yards. Dad always loved anything connected with the police department so I knew he would have loved it. For years I'd looked down on him for being such a straight arrow. I'd thought he was nothing but a sap.

But after Joe Testa's murder, I realized that Dad wasn't a sap after all. He was exactly what law enforcement needed: a good and decent man. The funny thing was, now, with my father gone, I didn't know a single one.

⊕ ⊕ ⊕

The headline read: SLAIN BUSINESSMAN LISTS 2 COPS IN WILL. When I saw that newspaper, I figured they might just as well draw a bull's-eye on my back. To my surprise, Joe Testa had named me and Jack Hinchy in his will, which made us both suspects in his murder—and potential targets for Marshall Caifano's muscle tactics.

Within days of Joe's death, I was questioned by ATF. I was also questioned by the FBI and the police department in Florida. And so were all of Joe's other heirs. But like I expected, nothing came of it. Joe Testa's murder was just one more "Killer Unknown" file.

After the investigation was over, the estate started going through probate, and all the relatives and heirs started feuding. Naturally, their biggest beef was over me and Hinchy. Joe had left us more than the family, so it was a real mess.

At first none of the Outfit guys had any idea what Joe's estate was worth. But they couldn't help but see all the publicity about me and Hinchy. It was a big story, a couple of cops getting a chunk of a "reputed" mobster's estate. It was all over the papers. Almost every day Sal would hit me with a bunch of questions. He was asking about Joe's holdings, his assets, how many heirs there were. Finally I asked him what his problem was.

He said, "Well, you know, that money's still gotta be paid back. Joe's dead, but that don't mean they're not expectin' you'll make things right. They're thinkin' it should come out of the estate."

"You gotta be shittin' me—"

"Hey, this deal with Marshall's not over," Sal said. "Not till it's straightened out. With Joe gone, it's you that's gotta get it straightened out . . . one way or another. You hear what I'm sayin'?"

I got his message loud and clear. Marshall was going to come after me next. I was pissed. At Sal. At Joe. At Marshall Caifano.

After that conversation, Sal was constantly on me about that money. He started feeding me all sorts of lines, working me real good. He'd say, "Listen, they're gettin' anxious. You gotta go see Marshall. You gotta talk to these people. You gotta be the peacemaker." Every day at least one conversation was about that money. Sal was like a dog with a bone. And just when I thought things were starting to die down, there'd be some article in the paper and there he'd go again. Naturally, when it finally came out that the estate was worth millions, Sal had a field day.

At first I thought Sal was trying to help me. But then I decided it was all just a ploy, that he was feeling me out. He was probably going right back to those guys and repeating every word I said. So the next time he started on me about the money, I said, "Jesus Christ, Sal. There's not any money yet. It's all on paper. There probably won't be anything out of that estate for years. I mean these fuckin' relatives are some real ball-busters. They're gonna take this deal to court. It's gonna be a hell of a mess. Me and Hinchy probably won't end up with a fuckin' dime when it's over."

Then Sal really laid it on me. He said, "Look, I understand that. I'm sure everybody else does, too, but you know, I got called in. Yeah, that's right, they called me in. The old man told me I gotta make you understand that this has to be taken care of. You gotta make this right, Mike . . . out of respect for Philly's family. I mean they got that money comin' to them. You know that. It's their dough. It wasn't Joe's. And Aiuppa and these other guys I been talkin' to—they don't think it's yours, either. You know what I'm sayin'?"

I told him I understood where he was coming from. I told him again how it could be years before the estate was settled. And then I stopped talking. I wouldn't talk about it. When he brought it up, I changed the subject.

Not long after that I started getting approached by some of the other guys. First it was Tony Accardo. Then, at a funeral, Joey Aiuppa called me into a room and asked me to sit down. He got right down to business. He said, "With Joe Testa gone, you're the one responsible now. It's your job to make sure that these guys get their money back. I told them you'd take care of it. Am I right?"

Naturally, I said, "Sure, Mr. Aiuppa." But that didn't mean I was going to come across with the money. I said what I had to. And then I got the hell out of there before I broke his neck with my bare hands.

I was beside myself. I was behind the eight ball all right, which is a place you never ever want to be with those guys. They had me—unless I got them first.

CHAPTER 27

lthough Charlie Nicosia didn't actually hold a job in local government at the time I knew him, he was a very well-connected political guy. He was extremely tight with Pat Marcy, the First Ward alderman and local political godfather, as well as Tony Accardo and Joey Aiuppa. Like Marcy, Nicosia's career had been promoted by the Outfit from the very beginning, particularly by Sam Giancana. Back in the late sixties, when the feds were investigating organized crime in Chicago, Nicosia was already making his way up the ladder, working as a clerk in the Illinois attorney general's office. Somewhere around that time, he took a pinch in a police raid along with Pat Marcy—who was then secretary to Alderman John D'Arco—and Tony Tisci, Sam Giancana's son-in-law and attorney.

After that brush with the law, Charlie Nicosia faded from public view. You didn't see his name in the newspaper. He wasn't listed on any FBI hoodlum list. You would've thought Charlie Nicosia was a nobody. But if you did, you would've been wrong.

By 1981, Charlie Nicosia was probably the main liaison between the Chicago mob and the politicians. When he talked, he could be speaking for Aiuppa and the Outfit or Pat Marcy and the Illinois political machine. Either way, you knew it was coming from the top. Charlie Nicosia was their mouthpiece.

To pull that off takes a very unusual type of personality. You've got to be a guy who can work both sides of the fence, and you've got to be able to shift from tough guy to back-slapping good old boy at any moment—sort of an iron fist in a velvet glove. That was Charlie Nicosia all over.

During the time I knew him, Charlie Nicosia was very involved in the court system, particularly civil court actions. He'd had all types of government jobs, so he knew everybody in the system, from the janitors and secretaries who had been around for years to the clerks, attorneys, and judges. Oddly enough, he was also real tight with all the head honchos up at the Mayo Clinic. Charlie got all the Outfit guys set up for annual physicals up there. Sal went. And I believe Carmen went, too. But not Hy Larner. There was no way Hy would set foot in the Mayo Clinic. One of his sons had had cancer, and he'd taken him there and they didn't save him. After that, Hy had no use for the Mayo Clinic. But Hy Larner was the exception. Charlie Nicosia got all the Outfit guys and their families going to the Mayo Clinic.

Whenever Charlie was up there, which was all the time, the switchboard was constantly announcing that they had a "call for Dr. Nicosia from Chicago," which is what everybody at the Mayo Clinic—the doctors, the staff, everybody—called him. And what the hell, they probably should've given him an honorary degree after all he'd done for the place. The guy sent hundreds of people up there. Of course, Charlie Nicosia wasn't any more a doctor than old Doc Rust had been. But unlike Doc, Charlie had all these plaques from the Mayo Clinic on his walls. They weren't medical degrees or anything like that—they were certificates he'd received in honor of his service to the clinic.

Most of the Outfit guys referred to him as "Dr. Nicosia"—partly because of his connection to the Mayo Clinic and partly because everybody knew you could go to Charlie when you had a problem and he'd "make it well." A guy with that type of reputation knew every fixer who ever was. So naturally, Charlie Nicosia knew Alan Masters.

Sal and I stopped by Charlie's house in Elmwood Park at least once a month with an envelope. Sometimes it would be just me making the delivery. While some guys would just take the envelope and shut the door, guys like Charlie Nicosia always wanted you to stay for a while, maybe have a cup of coffee.

Charlie had converted a four-flat building into a home. It didn't look like much on the outside, but once you got in the front door, it was like a palace. It was huge, decorated spectacularly with spiral staircases, Oriental rugs, Italian art—the works. You could easily sit twenty people at his dining room table. I always looked forward to stopping by.

I was making a drop one November afternoon in 1981, when out of the blue Charlie asked me if we could sit down for a few minutes. He said he had something he wanted to talk to me about. We went upstairs to his office, and he closed the door.

Charlie started telling me about a friend of his who had a problem with a book. It was a very important book. Charlie said that if it ever got in the wrong hands, a number of people would go down. It would mean some very serious problems for some very important people. Charlie said his friend had been having a little domestic problem with his wife, and one night he'd left the safe open and she'd taken that book. Now she was threatening to give it to the authorities.

"So as you can imagine, my friend's in a real bad spot here," Charlie said. "He needs some help. I'd like you to take care of this as a favor to me."

I was in a fucking corner. After my little meeting with Aiuppa about Joe's money, what was I going to say to Charlie Nicosia? I had no idea what he wanted me to do. I didn't even know who he was talking about. And that right there was the bad part of dealing with those guys. A favor with them can mean anything from whacking your best friend to carrying an envelope across the street. To them, it doesn't matter. A favor's a favor.

I said, playing it cool, "A favor? Sure. No problem. Anything you need, Charlie. We've known each other for twenty years. You know you can count on me. You just name it. Whatever it is, you know I'll handle it."

"Good," he said. "Then we can talk more openly now. My friend is Alan Masters. I've known Alan for a long, long time. Alan's done a lot of things for me, a lot of things for Eddie Vogel, too." Charlie paused. I knew what he was getting at by mentioning Vogel. He meant that book could bring down Hy Larner, too.

"If Alan doesn't get that book, a lot of my friends will be going away. We can't have that. So I want you to help get it. And I want both of them disposed of. Alan's book. And his wife."

I nodded. I wasn't surprised that Alan Masters had gotten himself in a fix like this. He bragged about being connected all the time. He'd probably told his wife way too much about his business, trying to look like a big man, and now he was screwed.

Charlie went on. "There's one other thing you gotta understand. No matter what, that broad is goin', but if you don't find that book, you gotta kill Alan, too. You understand? This is comin' from the top. We gotta get this handled."

I took a deep breath. This was not the type of thing I wanted to hear. I really didn't want to get involved with Alan Masters. Charlie Nicosia might've said he was looking for a favor, but he wasn't asking me—he was telling me.

When I left Charlie's home that day, he said he'd be in touch. He'd let me know what I had to do. So now, since I was "willing and able," it looked like all I had to do was wait.

⊕ ⊕ ⊕

My contact with Alan Masters was mainly business. If I wanted my money from all those referrals I was giving him, I had to go to his office and get it. I'd finally decided I didn't have to like the guy to do business with him. I'd make him money, and he'd make me money. And really, he was the best guy to deal with financially. If the referral was for an offense that would require a big payoff to the powers that be, I always gave it to Alan Masters. I knew he'd get the job done and take care of me. He was such a terrific fixer that you almost had to use him. He took notes, he kept files. He kept your deal on the square, with every penny accounted for. Alan was smart when it came to the green, but that's where it ended.

Everybody who knew Alan was aware of his problems with his wife, Dianne. The word on the street was she was nothing but a whore, but for whatever reason, Alan treated her like she was a princess.

Evidently he wasn't aware of her reputation. And I wasn't going to bring it to his attention. But of course, there's always somebody who will. One of the county coppers had told Alan about his wife, and that's when the trouble started.

It had been several weeks since I'd talked to Charlie Nicosia, but our conversation was still on my mind. I didn't say anything to Sal, or anybody else for that matter. I kept it to myself. But I was waiting to get that call from Charlie Nicosia. And now I was starting to dread it.

I was in Alan's office one day, picking up my envelope, when Alan started telling me about his wife screwing around on him. He wanted to know if I knew who it was she was messing with. He said he'd gotten a private detective named Ted Nykaza and another guy, a friend of his, Jack Bachman, to follow her around, but they kept losing her. He wanted to know if I'd put somebody from my security company, which did a lot of detective work, on her tail. He said he'd pay me very well. But I told him I didn't want to go that route.

Eventually he convinced me to follow her. Her routine was pretty straightforward. Generally, she went from the Moraine Valley Community College, where she was a member of the board, to this joint called Artie G's. Occasionally she might hit some other bars. I saw right away that she was doing plenty of fooling around, hooking up with a lot of different guys. After a couple of days I went back to Masters and broke the news. I wasn't brutal. I spared him the details. But even my watered-down version of Dianne's fun and games was too much for him. He looked at me across that desk and said, "I wanna do her."

I said, "What are you talkin' about?"

"I wanna kill her, that's what I fuckin' mean. She wants a divorce. She's threatening to take our daughter, Anndra, too. But there's no way she's taking my daughter, I'll tell you that fuckin' much. No way is that slut taking her. She told me she's gonna try to bankrupt me, too. She says she's gonna ruin me. But you know what? I'm not giving that bitch a fuckin' dime. I'm gonna do her."

I was just listening. I noticed Alan hadn't raised his voice once. He could be a pretty loud guy when he got going, which was something I never liked about him. But now he was real calm. Very low key. So I knew this wasn't just anger talking. He'd been thinking about this deal. And now he was plotting.

He took his key out of his pocket and shoved it across the desk toward me. "Here's how I want it done. You take the key to my house. I'll make sure Anndra won't be there. You make it look like a home invasion. Take all Dianne's jewelry. I'll tell you where it's at. Beat her, rape her . . . do whatever the fuck you wanna do, I don't care. Just kill her in the fuckin' house and leave her. I'll have an alibi. You just take that key." He motioned to the one in front of me. I was in a bind. I was wondering if this was Charlie Nicosia's deal or if Alan was playing it on his own. There was no way to know for sure, so I took the key.

After a week or so I decided that since I hadn't heard from Charlie, this had to be Alan's personal game plan. So I sat tight. I didn't do a thing. I had that goddamned key for two months. Alan was calling all the time. He'd say, "Listen, she's gonna be home later, it's all set up." I always begged off. I was doing this, I was doing that. Finally I decided I wasn't doing shit for Alan Masters. I hadn't gotten a call from Charlie. So I just said to hell with it.

I went over to his house and handed him the key. I told him, "Get somebody else. There's a lot of things I'll do, but killin' your wife's not one of 'em."

Alan said he understood, but he didn't let me off the hook completely. He said he'd come up with a better idea. He said she was working downtown for a guy named Gerry Cosentino who was running for state treasurer. He figured that when she walked up the stairs to get to the office, I could kill her there in the stairwell and make it look like a robbery.

He was pushing me very hard. Still, I was not at all interested in doing this deal. I said, "Look, I'm not doin' it, okay? Just forget about it. Just count me out of the fuckin' deal."

So now I was out. But I also had knowledge that this guy was planning to kill his wife. There are a lot of men who might talk about doing their old lady. They eventually come to their senses or they kill them. One of the two. I was hoping the whole thing would just go away.

A few months went by and I was in Alan's office, and he started pulling out these love letters that Dianne was getting from this guy. Alan started reading these letters out loud. After he finished, he asked me if I had somebody who could wire phones. He said he'd been out of town a lot and he thought Dianne was talking to this guy on the phone. I told him no problem.

I got my phone guy out there, a former county copper who was working as my deputy chief, and we did Alan's phones. We put in voice-activated tape players with about four hours of tape. A few days went by and we got the tapes. I didn't listen to them before I took them up to Alan's office and played them for him. I could not believe it—they were nothing but her talking to a girlfriend about having sex with this swinging dick. They were very graphic. I was watching Alan's face. It was killing him. I told Alan that was enough, that he didn't need to hear any more, but he insisted on hearing every last word. It was really terrible. She described how this guy poured wine on her pussy and licked it off and then put the wine bottle up her snatch.

That was just one of the tapes. On another one she was talking about how she was screwing this guy in the living room when Alan's three-year-old daughter walked in. She was asking her girlfriend if she thought a kid that age would remember it, if maybe she'd think it was a dream. She was worried Anndra would tell Alan.

When Alan heard about his kid, he lost it. He exploded and

went crazy nuts. He started screaming, "You know what I'm gonna do? You know what I'm gonna do?"

I knew all right. And there was nothing I could do to stop it.

⊕ ⊕ ⊕

When I got the call from Charlie Nicosia saying he wanted to meet me at a pancake house, I knew things had progressed to the next level. We didn't even look at the menu. We ordered coffee, and Charlie got right down to it. He told me Alan's situation was much worse than it was the last time we'd talked. His wife still had the book—although Alan hadn't mentioned anything about a book to me—and she was giving every indication that she was going to use it against him when she filed for divorce, which could be at any time. Of course, Charlie said, making the information in the book public wouldn't just hurt Alan Masters; he reminded me that there were a lot of very important people on the line as well.

I figured Charlie was going to bring up that favor he wanted me to do for him. But instead he said, "I understand you don't want to do the job, Mike. And that's okay. We've got somebody else who's going to take care of that part."

He went on to tell me that it was going to look like a disappearance, like she'd run out on Alan—which was totally believable given their current marital situation. Then Charlie added, "There is one small detail, a very important detail, that we need handled, Mike. So I still have a favor to ask. When the time comes, I want you to get rid of her car. But right before you do that, I want you to make sure that book isn't in there. If it is, get it out and bring it to me right away. As far as the car's concerned, dispose of it any way you want. Just as long as it's never found. That's all I care about. With her in it . . . or, I don't know for sure yet, there may be two of them. Don't worry about that part, just get rid of the car. Make sure nobody ever finds the fuckin' car."

So that was the game plan. At least my part of it. And as I found out later, Alan had a bunch of red herrings made up to try to make her disappearance look good—sightings of her in California, that kind of thing. Really, I didn't want to hear about it. I just wanted it over and done with as fast as possible. As far as I was concerned, I was following one guy's orders: Charlie Nicosia. I'd do my part and I was out of it.

They set a date to do the job. March 2. It was a week away. We worked out a signal so Alan would know when I'd disposed of the car. I was to call his house and let it ring one time and hang up.

But the night it's all supposed to come down, I go to the Blue Front Lounge on Archer, where the car was to be delivered, and the joint's closed. The parking lot's totally empty except for one white and brown Jeep, which I figured belonged to the shooter. I figured whoever was supposed to deliver her car had to use that vehicle to leave the scene. He wasn't just going to stroll down the highway. So he'd left his Jeep there earlier. Maybe Alan had picked him up. I took the license number and ran it later. It came back registered to the father of a guy I knew.

I sat there in my car in the bushes until two o'clock in the morning. At that point I saw a guy walking down the road. I watched him get in the Jeep and leave. I recognized him. I'd seen him before, I just couldn't think of his name.

As it turned out, Alan's wife didn't show up where they'd thought she would. Instead she went with her boyfriend to a motel. So now everything was on hold. It was agony. I wanted this deal over—or forgotten—in the worst way.

Several weeks went by, and I got the call again. It was supposed to go down on March 19. That night I went back to the Blue Front, and this time it was open for business. Sure enough, there was that same white and brown Jeep. Same plate. So now I was almost positive who was involved.

It was after two o'clock, the place was still open, when I saw Dianne's Cadillac heading my way. They parked it around back, and I saw the guy get out and pitch the keys under the floor mat, just like he was supposed to. I waited until he left before I went over. And that's when I saw that the right rear tire was flat. I could've gone nuts right then and there. Didn't the dumb son of a bitch see that he had a flat? Or maybe he just threw the hot potato my way and said, Good luck, sucker.

What was I supposed to do? Drive around with a flat tire and a couple of bodies in the trunk? To make matters worse, the Cadillac was ugly as sin. White with a bright yellow top. Me and that car were going to stick out like a sore thumb. I did not want to be seen driving Dianne Masters's car the night of her disappearance.

Originally I'd planned on zipping right over to the canal and dumping the car. I figured the entire process might take five minutes. But with that flat, forget about zipping; I'd be lucky to make it there at all. And from the looks of things I'd practically be driving on the rim, too, so not only would it be slow going, it would be noisy—not a good combination when you're trying to sneak around. And there I was, the chief of police on top of it. People knew me, for Christ's sake.

I was sweating pretty good now. I checked my police radio and heard that our cars were headed over to a restaurant in Bridgeview for a shift change, which meant there wouldn't be anybody else from the force in my area for a while. It would be tight, but I probably had just enough time to get to the canal, which was about half a mile away. I got in the car and headed for the canal. Once I arrived, I looked through the car for that book Charlie Nicosia had talked about. There was nothing. Not even a Kleenex and I wasn't about to open the trunk and risk leaving anything that could lead to me. I put all the windows down, so when the car went in it would fill up and go down in a hurry. Then I put it in DRIVE, and over it went.

It was real nerve-racking watching the car go down. In my rush to get rid of the damned thing, I'd forgotten to turn off the headlights. It seemed like it took forever for those lights to finally go off. It was eerie watching them go dimmer and dimmer, until they finally faded away and the water turned black again. You could say I was relieved.

But as far as what had happened to Dianne Masters, I didn't feel much of anything. Dianne was very calculating, but she played the wrong game. When she stole Alan's book and threatened to use it, she unknowingly got the Outfit involved. And that's when she crossed the line. She didn't know who she was dealing with then.

⊕ ⊕ ⊕

When you're a cop, you know how to shake off everything that could possibly lead to you or connect you to a crime. I didn't carry a wallet that night at the canal—I don't know how many times a wallet's fallen out of a guy's pocket and it was "case closed."

I'd worn an old coat, a stocking cap, cotton gloves, an old pair of jeans, and a pair of worn-out tennis shoes. And of course, I pitched it all right away. I wanted to distance myself from that Masters deal as far as possible.

But Alan Masters wasn't going to let that happen. About five days after Dianne's disappearance, I finally saw him. He was in his Corvette, with his daughter, and he came up behind me and flashed his headlights like he wanted me to pull over. When I stopped, he got out of his car and ran over to my window. He was real worked up.

"Thank God Keating's got it all handled," he blurted out. "Let me tell you, it's been wild. You won't believe it," he said. "You know, that motherfucker wouldn't shoot her. He backed out—"

Was the guy nuts or what? I cut him off. "Shut up, Alan."

But he wasn't listening. "Yeah, the cocksucker refused to shoot the bitch—"

"Shut up, Alan. Shut up and get in. Now."

He ran around and got in. "Yeah, he backed out . . . can you believe it? He'd already hit her with the tire iron. She went down when he did that. It made a real bad sound, like he crushed her skull. I figured she was dead. But I wasn't sure. He had a fuckin' gun. He was supposed to finish her off. I told him to shoot her, but he wouldn't do it, Mike. We were outside. It happened between the car and her getting in the house. Now I was worried. What if she wasn't dead? Okay so he hit her . . . but then I heard her make a couple of sounds, and I told him he had to shoot her. He had the gun, but he wouldn't shoot her. So finally I took the gun away from him and I shot her. I shot her twice, Mike. I did her myself. Right in the fuckin' driveway." He smiled. "Can you believe it?"

Looking at Alan Masters sitting there next to me in that car—well, I was wondering if me and Charlie Nicosia were going to have a major problem on our hands. The guy was coming unglued. "You need to relax, Alan. Go home, be with your kid. Don't think about it for a while. Think about Anndra. She needs you now. Look, she's over there in the car waiting for you. Go on home. Relax."

He was very fidgety. "You're right," he said, and he started to get out of the car. But then he stopped and leaned back inside and grinned. "So tell me, what did you do with her? Come on, I won't tell anybody."

"That's none of your fuckin' business, Alan. You hear me?"

"Hey, I think I have a right to know. . . . I have a right to know what you did with my fuckin' wife."

I wanted to pop the dumb son of a bitch right there. What the hell was wrong with the guy? I told him never to ask me that question again. Ever.

After that, Alan Masters didn't have much to say when he was around me. Which was just fine. I wanted to forget about the entire thing and move on with my life. I had Joe Testa's estate coming through. I probably had three or four million stashed away on top of that. I didn't have to work another day in my life if I didn't want to. You could say it looked like I had it made. That's what it looked like anyway.

CHAPTER 28

I probably should've left the country. But of course, I didn't. I guess what made me stay, what influenced me more than anything, was the endless stream of green. Sure, I had enough, but that wasn't the point. It was the thrill of it, of seeing all that dough come through my hands every single day. There were times when I had two and three duffel bags full of hundred-dollar bills in the trunk of my car. It wasn't all mine, but a piece of it was. The best part was, it never dried up—that dough just kept right on coming. You need a few bucks, you've got it.

How could I turn my back on all that money? And what about the action? I was just as hooked on the rush I got making all that dough as anything. It gets to be a high, doing deals all day long. As a cop I was rooting. As an Outfit guy I was skimming. Laying on a beach in Panama was okay now and then, but the truth was, more than forty-eight hours of that crap and I was bored out of my mind. No, I needed to stay in Chicago, for better or worse.

When Alan told me Keating had the situation with Dianne's disappearance handled, I figured things would be better than worse. With Keating in charge, the county would treat the case like it was just another runaway. Believe me, you can drag that type of investigation on forever and ever—until nobody even remembers who they're looking for. That's what I figured would happen. I think we all did. Nobody counted on Dianne Masters's brother, Randy Turner, getting so involved.

Turner was on Alan from day one. He met with the detectives on the case and insinuated that Alan was involved in Dianne's disappearance. He was making a lot of noise, demanding the police contact the media. Keating managed to stonewall Turner for about a week, but then the story broke. When reporters rushed to get statements from Dianne's family and friends, Alan offered a ten-thousand-dollar reward but refused to comment. Meanwhile Randy Turner had a field day. It seemed like he was giving interviews for weeks. Still, I'd seen how the press handled sensational cases like this one a thousand times before. The twelve disciples could walk down Michigan Avenue and it would only make the front page for a day. Next thing you knew, those reporters would be on to something else. I knew things would die down eventually. I was just hoping it would happen sooner than later. More than anything, I wanted to forget the entire ordeal. Ever since that night at the canal, the image of those headlights slowly dimming beneath the water had been on my mind. I couldn't seem to shake it. And I had enough on my mind without worrying about what Alan Masters and his pals had done to Dianne.

For starters, my personal life was in complete chaos. After seven years, my marriage to Janice was finally coming to an end. Counseling had been out of the question. There were so many reasons it didn't work I wouldn't have known where to begin. All I knew was that our marriage had been a big mistake. We were just too different. She came from a classy southern family. Me, I grew up in a blue-collar neighborhood wearing my sisters' blouses. Janice expected we'd have the finest of everything—and I really tried not to let her down. She had the grand piano, the crystal, the silver, the china. But I needed that junk like I needed another asshole. And really, it was way out of my league.

Of course, that was only part of the problem. I'd definitely made things worse between us; when I wasn't traveling, I was working at all hours. I hadn't been any kind of husband. She never knew what the hell I was doing or who I was doing it with. And she'd learned better than to ask, too. I think we both realized there was nothing between us anymore. Still, Janice took our divorce real hard. As for me, I just wanted out, so I gave her a real sweet deal, including alimony and all house-related expenses until the divorce was final. I was also responsible for child support—for our son, Michael, and her daughter from a previous marriage, who I had adopted. Like all divorces with kids involved, it was a sad—and expensive—proposition.

In the divorce settlement, Janice ended up with the house, but in no time at all she put it up for sale. It was a beautiful home, with a pool, on an acre in the woods. It was the only house I'd ever built, and it was on the lot that Sam had paid for. Call me a sentimental fool, but at that point in my life, money was not an obstacle. I bought the house from Janice and moved back in.

Naturally, it wasn't long after my divorce that I was right in the middle of new love affair. Sherry was a very pretty, very classy woman. It could have been déjà vu all over again, but this time I felt things were different. For most of my life, everything had been compartmentalized. I had my wife and family on the one side and business on the other. For all the highs, I was often very lonely. Every woman I'd ever been with was completely in the dark about what my life was about. They were also very naive. They didn't fit in with the other Outfit guys' wives, either, which had made it difficult for me to socialize with Sal and the rest of his crew. But right from the start, Sherry hit it off with everybody. And far from being naive, she was extremely street smart. She knew I was friendly with the Bastone crew and what being a cop in Chicago was about. She also knew what I'd been up against ever since Judge Wosik's decision had reinstated me as chief of police in Willow Springs. After a decade or so of wrestling with the village board, it wasn't fun for me anymore. When the village board filed an appeal with the Illinois Supreme Court, it was obvious things weren't going to get any better, and I started thinking about getting out of law enforcement altogether. That didn't matter to Sherry. "I'm not in love with a

cop," she told me. "I'm in love with you." Sherry was behind me all
the way.

The case before the Illinois Supreme Court was scheduled for
June. And every day until then seemed like an eternity. Just being at
the station was a struggle. Mayor Militello was busting my chops,
and I was busting his. Day after day it was going on like that. It got
so bad that I didn't go in half the time; I'd have a couple of my cap-
tains run the department while I was running around with Sal, traveling to Vegas, to Miami, and Panama. Believe it or not, the entire
time, nobody in the village knew where the hell I was. The mayor
never even knew I was gone.

With summer approaching, my career in law enforcement was
hanging in the balance. I'd been a cop for almost twenty years, and
I'd been under a lot of scrutiny during that time. There'd been a ton
of investigations. There'd been inquests into shootings I'd done in
the line of duty. A grand jury investigation into my selling badges.
The big stink over my security company. The time the IRS came into
the station and confiscated all my records. None of it amounted to
anything. But the pressure was constant.

With the lapse in high-profile investigations, the local papers
took over. By June they were full of allegations—that's all they
were—about corruption in the police department. But my problems
with the village board didn't really have anything to do with corruption. They just wanted to make it look that way. It was all
strictly political. But of course, in Chicago, politics and corruption
go together like ham and eggs.

I knew in my heart I'd done a good job for Willow Springs. For
years, I'd basically run that little community. We didn't have any
crime. And I was proud of that fact. We had no burglaries, no
armed robberies, no assaults. Before I became chief, we'd had a rash
of auto thefts, but once I got in they stopped. I just went to the chop
shops guys and told them to knock it off, which they did out of
respect for me.

We also didn't have a single murder the entire time I was out
there. Sure, we found a lot of bodies—but they weren't ours. His-

torically, Willow Springs was a great place to dispose of a body, thanks to its woodlands and out-of-the-way spots. But none of our residents was killed during my tenure; any body we found was an individual from outside the area who'd been murdered elsewhere and brought to Willow Springs and dumped.

I figured a record like that had to count for something. I had most of the community behind me. But in the end, none of that mattered. When the Illinois Supreme Court agreed to hear the case that June, I knew it was over. I walked into my office and cleaned out my desk.

⊕ ⊕ ⊕

Meanwhile, just as I'd expected, most of the publicity surrounding the disappearance of Dianne Masters died down. By the time summer rolled around, I was relieved to find I wasn't haunted by it so much anymore. When it did come up, I reminded myself that killing Dianne Masters hadn't been my idea. I'd gotten involved thanks to Charlie Nicosia. He'd asked me for a favor, and what was I going to do? I'd been around long enough to know it wasn't smart to turn down the Outfit when they called in a marker. I'd been in a very tough spot. Besides, all I'd done was dump a car. Really, I hadn't even known for sure if there was a body in that trunk. And as far as Alan Masters and Jim Keating were concerned, they hadn't been my pals before and they weren't now. I was off the force, so I didn't even have to conduct business with Alan anymore.

That summer I spent most of my time with Sal Bastone, working with Hy's crew, or traveling. For now, the subject of the money I was to receive from Joe Testa's estate had been dropped. The pressure of me being chief was gone. The IRS was clearing me in its latest investigation. My marriage to Janice was over. And although I had no intention of ever remarrying, I was starting a new life of sorts with Sherry, who had moved into my home and set up housekeeping. You could say I was wiping the slate clean.

In August, Sherry and I attended a big dinner and reception held in honor of my "outstanding community service" as chief of police. Thrown by the Willow Springs police force, it was part retirement party and part fund-raiser, intended to pay for the legal expenses I'd incurred during my fight with the village board. Seven hundred people were in attendance. It was all the big shots and top politicians from all over Chicago. I walked around that room the entire evening, shaking hands and thanking people. I felt like I was on the verge of starting a whole new life; everywhere I looked, there was opportunity smiling back at me. The truth was, any one of those people could be my next break, my next big deal.

It would've been the perfect ending to my old life—you know, the part where the hero gets on that white horse and heads off into the sunset—except for one thing. It wasn't over.

⊕ ⊕ ⊕

In September, a Willow Springs cop was out by the canal and happened to notice some tire tracks leading over the edge into the water. Concerned, the police department called in a couple of scuba divers, Tom Skrobot and Gerry Los. Skrobot and Los hadn't been in the water five minutes before they announced they'd made a surprising discovery: the bottom of what was formally known as the Chicago Sanitary and Ship Canal was covered with abandoned cars. That wasn't any surprise to me. And it shouldn't have been to anybody else in Willow Springs, either, because a large percentage of the populace had used that canal as a dumping ground for years. In the past, when somebody pulled a car out of there, nothing ever came of it, nobody ever got pinched. But this time I'd heard that things were going to be different. Supposedly, my old pal Mayor Militello and his Keystone Kops were going to make a big production out of the deal. Suddenly, all those old images of that night at the canal came flooding back. Within days, Willow Springs had launched a high-profile investigation into dumping in the canal, bringing in a crane and putting the divers back to work.

Just as the authorities announced they'd pulled their first car out of the canal, Sherry made an announcement of her own: she was almost four months pregnant. Her timing could not have been worse. To say I wasn't thrilled would be a huge understatement. I had no desire to get married again, and I definitely didn't want to get married under those circumstances. Sherry and I liked each other and we had a good time together, but that was the extent of it. I'd never considered the idea of "forever" with Sherry. But now it looked like I'd have to. I was always a sucker for the old do-the-right-thing routine; we set a wedding date for December.

Between my upcoming marriage to Sherry and all the attention the press was giving the investigation into the canal, I was very uptight. And as the days went by, my anxiety continued to grow. The word *dread* is probably the best description of what I was going through at that time. Every day it was another car, another make, another model. Every day I waited. And every day I wondered, with an increasing sense of dread, would this one be a Cadillac? For weeks, day after day, it went on like that. I couldn't sleep. I couldn't eat. I couldn't talk to anyone about it, either. I'm sure Sherry thought it was our upcoming wedding—along with having a baby due in May—that had me so down. Of course, I couldn't tell her any different; I was totally on my own. It was excruciating.

During this time, I also started worrying about what Charlie Nicosia would have to say when they pulled Dianne Masters's Cadillac out of the canal. He'd given me a job to do—not that I'd had much of a choice—and now look what happened. Plenty of guys had been killed for less.

After two months the authorities had recovered around twenty vehicles, all of them connected to some type of insurance fraud. And still no Cadillac. The papers were calling the canal "a grave-yard of stolen and abandoned vehicles." I felt like it was only a matter of time before the whole world knew that it was much more than that.

On December 11, 1982, it was all over the news that they'd pulled a white Cadillac with a yellow vinyl top bearing license plate number DGM 19 out of the canal. In the trunk of the car, they'd found the badly decomposed remains of a woman, her skull fractured into over forty pieces. Later an autopsy would show evidence of blunt force trauma to the skull, with cause of death determined to be two bullet wounds to the head. The body was identified as that of Dianne Masters. Her death was ruled a homicide.

I guess I was almost relieved when they found that car. Finally the waiting was over. But now there was the very real possibility that I'd feel some major repercussions. For one thing, there were Charlie Nicosia and the Outfit to worry about. For another, Dianne Masters's disappearance had been reclassified as a murder. So in reality, things were far from over.

But the more I thought about it, the more confident I became. There was no evidence that tied me to the murder. All I'd done was drive a Cadillac half a mile and shove it into a canal. There'd been no witnesses. I hadn't left any fingerprints, and I hadn't opened the trunk. I also hadn't killed her. As far as Charlie Nicosia was concerned, I knew he wouldn't be thrilled that Dianne's car, complete with her body in its trunk, had suddenly appeared on the scene. But I also knew that from Charlie's perspective this had been strictly business. Since there was nothing connecting him to Dianne's murder, I had to believe he would handle the situation in typical Outfit style: he'd wash his hands of the deal and let Alan Masters and his shooter take the fall. Then we'd both be in the clear.

CHAPTER 29

After I left the force, I spent all my time with Sal. Sometimes we were together twenty hours a day, every day. I continued to operate Swift Security, and I got a job as deputy sheriff with the circuit court clerk in Cook County. It was a "Casper" job—meaning no-show—that put me on the payroll for about five hundred bucks a week. But it wasn't the money I cared about. That job meant I was a deputy. So now I had a star again. And a weapon.

Sal had always gotten off on the idea of me carrying a gun. He loved the fact that I was able to conceal a deadly weapon and do it *legally*. I never understood what difference it could possibly make to him, because most of the time Sal just went from restaurant to restaurant all day long. And if he wasn't doing that, he was making calls from pay phones. Sal Bastone wasn't exactly living on the edge. In fact, in some respects what Sal did for a living was boring. There were days when we'd meet ten guys at ten different restaurants and I'd hear Sal tell the same joke ten times. But then there were the times when Sal got off the phone and said, "Let's take a ride," and we'd go by some guy's house and the garage would be full of swag. There might be refrigerators and appliances. Or maybe it was shirts. We always walked out with our arms full of stuff. So in some ways, working with Sal was also like Christmas every day.

Sal knew the car thieves, the hijackers, the burglars. But don't get me wrong, it wasn't like Sal Bastone needed a handout. The guy had a ton of money. He just didn't have anyplace to spend it. The Bastone crew had the arm on everybody; everything was always on the house when Sal walked in the door. We'd waltz into liquor stores, clothing stores, butcher shops, and we'd carry stuff past the register and out the door.

There was one store that was owned by two Jewish brothers who had been on the arm for juice money forever. They were real big gamblers. They owed Sal so much money that they couldn't pay it back in a million years. So every so often, Sal would bring four or five guys in there and just tell them to take whatever they wanted. I'd pick up suits, shirts, socks, shoes, then I'd put them in a bag and go to the cash register, and those two brothers would just smile and say, "Thank you."

Sal had car dealers on the arm like that, too. If we needed a car for a while, he'd just walk in a showroom, point to a car, and tell them to have it ready in twenty minutes.

Now when all your dough is coming in and none of it's going out, it's hard to spend it fast enough. The average person wouldn't call that a problem. But it was for Sal. He had so much money he didn't know what to do with it. He was bringing in at least two hundred thousand a month on the skim, all cash. Out of concern for the IRS, he couldn't stick it in banks or savings and loans, so he was constantly looking for ways to hide it in an investment. He put money into restaurants that were in different people's names. He invested in harebrained schemes that never went anywhere. He bought his wife and kids anything they wanted. They all traveled around the world.

Sal bought homes all over the country. He bought homes for his relatives and for the friends of his relatives. He had three homes in Wisconsin alone—one of them Hy and the Bastones used for their secret meetings with Aiuppa or Lansky or one of the other top bosses. For Sal, it was just buy, buy, buy.

But that wasn't enough. Sal always thought his brother Carmen had more. He was very jealous of Carmen. It was always Carmen was jetting around, Carmen had all these broads, Carmen this and Carmen that. And it didn't help that, unlike Sal, Carmen was very outgoing as well as trim and good-looking.

Sal had been a handsome guy when he was young, but over the years he'd gotten big and blown up. Sal just couldn't help himself around food. There'd be some weeks when we'd stop at a dozen restaurants and he'd have us eating at every one of them. The next week he'd be down in the dumps about his weight and he'd go on a diet. One time he signed up at a clinic for a weight-loss program, and every three days we'd have to stop there so he could go in and weigh himself. He'd come out shaking his head. He'd have these cans of malt crap he was supposed to drink for the next three days.

One day we went into a restaurant and I ordered a big juicy steak, but Sal couldn't have anything on the menu. He had to drink that malt shake, and consequently he was just miserable. He said, "So how do I look? Tell me the truth . . . do I look like I'm losin' weight?" Before I could answer, he shook his head and sighed. "Yeah, I know. I still look fat as a pig."

Really, it was pathetic, seeing a man with the power Sal Bastone had acting like that. Here was a guy who controlled millions of dollars, who had the power of life and death over people, and he was still just as screwed up about his looks as any other middle-aged man.

One morning I went to pick him up and he came out of the house looking very upset. At first I thought Hy had gotten hit. Sal got in the car and said, "You're never gonna believe what happened. My goddamned hair's fallin' out. That's right, it's fuckin' fallin' out. It's from this fuckin' diet, too, I know it. I'm goin' bald because of that diet. I'll have that fuckin' doctor whacked if I go bald because of him and his fuckin' diet." Sal was so upset that we had to go right then to see another doctor. Of course, they took him off that weight-loss plan right away. Sal was terrified of being fat, but he was even more terrified of being bald.

Finally Charlie Nicosia got Sal into the Mayo Clinic, where he had his stomach stapled. Sal lost a lot of weight after that and had to get a whole new wardrobe. He looked real good, too. Some people thought he even put Carmen in the shade. But Sal was still jealous of Carmen. He always thought Hy Larner favored his brother. The funny part about it was that Angelo thought Hy favored both of them over him. But really, it wasn't that at all. It was just that Hy could see what a guy's strengths were and where they'd do the best job for the organization.

Carmen was a "people person." He was a real wheeler-dealer and very social, so he belonged in a line of work where he was massaging the big shots. His forte was handling political payoffs and all the maneuvering that went on overseas. At that time Carmen had completely organized Spain for Chicago.

Angelo, who'd been a bodyguard and a cop and knew how to handle himself, was completely different from Sal and Carmen. When he came back to Chicago from Vegas, Hy had him run all the machines in the Heights, a real rough area. Angelo had a route crew and a collection crew that worked a huge territory from Blue Island all the way to Indiana. There was always somebody robbing the machines or holding up the route man in the Heights. Nobody even wanted to drive through there. It was all hillbillies and crackers and a bunch of lowlifes. But Hy knew that Angelo would be able to handle it.

Sal was the "number cruncher." He was serious and very analytical, giving him expertise in handling the money. Hy had Sal oversee all the skim that came into Chicago as well as all the money that came from the machine operations like Zenith and Rentall. When you're moving that much dough, a person like Sal Bastone is critical to the success of the operation. It was a hell of a responsibility.

⊕ ⊕ ⊕

Zenith Vending operated within the law—as much as you could in the machine business. Zenith didn't just repair machines. If Hy came up with some new idea, Zenith might actually build one, too. In fact, Zenith's electronic technicians built the very first video poker machines. Supposedly, there was no such thing until Hy

Larner came up with the idea. Sal told me that Hy cut a deal with Bally for the use of the technology. When Hy showed them what big moneymakers video poker machines were, Bally put them in all its casinos. By 1983 the biggest moneymaker in the casino industry was video poker and I believe it's still right up there.

Unlike Zenith, Rentall drop-shipped its machines and parts. It didn't keep much in stock. It got an order from a customer, and when the order came through from the manufacturer, Rentall shipped it out. Rentall bought machines from all over, but like all of Hy Larner's other operations, it was doing business with Bally—it was just under the table or through the back door.

Sal had one of his buddies, Joey DeVito, in charge of Rentall. It was a major position because Joey was the guy everybody turned their money over to. He kept all the different sets of books. As far as I know, Joey DeVito still runs Rentall, since they moved the company within a mile or so of Bally Manufacturing, in Bensonville, Illinois.

Rentall was basically a middleman. It had a rental agreement with the operators. That way nobody could ever come back and say it was operating gambling machines. What an establishment did with the machines on its premises was its business. All the stuff that was illegal was handled by Hy's other companies, like Flip Amusement.

Between Rentall and Zenith alone, Sal was overseeing several million dollars a month in skim. The money from all the domestic operations came in on a monthly basis. To count it, we'd get a suite at the Hyatt Regency or the Oak Brook or Rosemont Hyatt and set up a counting room.

We'd start at around two or three o'clock in the afternoon and go on like that for two whole days. The route guys always went to the currency exchanges or the banks first and got rid of all the change and small bills. Sal didn't want it bulky, so most of it was in hundreds because they were easier to conceal.

There were thirty or more route guys, and once they got the word, they'd start dropping off their money in old green army duffel bags. They didn't get a receipt; there were never any records of their transactions. The only thing Rentall reported to the IRS was how much it paid the route guys.

After all the money was delivered, we'd lock the doors and start our count. It was a hell of an operation. We'd count the money three or four times and put it into ten-thousand-dollar stacks on the bed. Then we'd rubber-band each stack and organize them into piles. You could usually tell from the size of the pile who it was going to—the bigger the pile, the higher up the guy was in the organization.

Aside from overseeing the count, Sal also organized the distribution. With the domestic money, there were at least seventy-five people getting an end, from Wisconsin to Indiana: politicians, policemen, mob guys, and fixers like Pat Marcy. We'd have four, sometimes five men making deliveries, including me and Sal.

We kept as many as twenty of the deliveries for ourselves; Sal always took the heavy hitters, so he could be seen as the main man. We'd go nonstop for two solid days, no sleep, no going home. Once we started, we didn't stop. Sal wanted that money out of the car as fast as possible. He didn't want to get caught with half a million bucks—cash—in the backseat.

We'd start our rounds in the subdivision of Oak Brook, at Joey Aiuppa's house. We'd have to spend half an hour there, which I didn't enjoy at all. I always had the deal with Joe Testa's money in the back of mind. But whether you liked it or not, when you went to the old guys' houses you had to sit down and have a cup of coffee, maybe even a bite to eat. It was a social thing with them.

After Aiuppa, we'd meet with Tony Accardo's driver and bodyguard, Sam Carlicci. We'd give him the package for the old man and head downtown to take care of another top dog, Gus Alex, who continued to maintain his position as political fixer and North Side boss. And of course, Pat Marcy was on our delivery list. Even though this deal had nothing to do with the Outfit's traditional payoffs to the cops and politicians, Marcy always got a package off the machines, generally around thirty or forty grand a month. Later in the month Marcy got another hit when we delivered the regular envelopes.

As for me, I always got a cut from Sal. He took real good care of me. During a count, he'd throw me a stack and I'd put it in my coat pocket. I'd just made ten grand. No big deal. And not bad for two days' work, either. Not bad at all.

⊕ ⊕ ⊕

On top of all the dough we had coming in domestically, there was also a ton of money coming in from the operations in Central and South America, the Philippines, Spain, and the Middle East. Messengers came in once a week from all over the world with suitcases full of cash. Generally they came in on a commercial flight through O'Hare, Palwaukee, or Midway. But of course, Hy also had his own fleet of jet planes out of Miami and Panama. Sometimes I'd go out to O'Hare or Palwaukee and meet one of them. Usually one of the crew would hand me two steel suitcases full of money and off they'd go.

Hy Larner had a huge network of bagmen. In a five-year period, I must have made over three hundred trips to O'Hare alone, picking up couriers. There was constant concern that we might have a robbery, so I usually had another car, with some muscle, behind me. I'd pick up the courier, and we'd head for Bucky Ortenzi's place in Melrose Park.

When it came to how the couriers themselves handled the money they were bringing to the United States, they were all different—mostly because some countries were more difficult to maneuver through than others. The U.S. military bases and Panama were easy; they used U.S. money in their gambling operations, so getting the cash out of the country didn't require much hassling with exchanges. But countries like Guatemala—where Hy's cousin and lawyer Harold Lee had an operation—had a lot of problems. All the machines in Guatemala used Guatemalan money, no U.S. currency. So Harold Lee's guys had to count it and take it to a bank down there and have it converted into U.S. dollars before they could send it to New Orleans, Miami, or Chicago. And that could be a major pain in the neck in a poor country. A bank might not have a million bucks on hand to swap.

When Chicago still had Iran, they had a hell of a time converting money over there. And it was the same with the Philippines. To bring the money into the States from the Philippines, they had to leave Manila and fly to London, where they went to the Bank of England and changed it to U.S. dollars. Then they'd bring it to Chicago on commercial or private flights. Either way, they never had a problem with customs.

There were times when a courier would have two or three big packs of Japanese yen or German deutsche marks because an exchange didn't have enough U.S. dollars to make the swap. They'd have to bring it to Chicago and Sal would send somebody to a particular currency exchange we always used on the North Side. Occasionally we had couriers come in with these two-ton suitcases because an overseas exchange hadn't had enough large denominations for the swap. Their suitcases would be stuffed with two million dollars in twenties and tens. A valise full of hundreds isn't that heavy, but try it with small bills sometime; it's a backbreaker.

Besides bringing money into Chicago, Sal made me understand that Hy had just as much coming in through New Orleans, Miami, Salt Lake City, and Las Vegas. I just never saw any of it firsthand, although I did continue to courier funds myself through Las Vegas. Of course, I saw how the money was handled after it got to Chicago. Sal always took a few stacks before we arrived at Bucky's. He was probably skimming something like thirty, forty thousand a month off the overseas operations alone.

Once we arrived at Bucky's, we carried the suitcases in and Bucky took his cut. Then everybody put on rubber gloves. Those were Hy's orders. Rubber gloves, no fingerprints. After we got our gloves on, we'd take out all the money and put it on the conference table in Bucky's dining room. Because of all the currency exchanges, everything was already counted, so we'd stack it up on the table in fifty-thousand-dollar bundles.

From Bucky's, the money went to a condo that Sal bought specifically for the purpose of dividing up and preparing the money for delivery. Sal was in charge of organizing and distributing the international funds as well as the domestic money. He would call out, "Put forty thousand in there, put eighty in there, fifty thousand in there, seventy thousand . . ." and so on.

Everything was separated into green garbage bags. We'd put the money in the bag and then roll the bag around it real tight, into a five-inch-thick brick, and then we'd put masking tape around the center and over the ends. Somebody put initials or a nickname or some code word on the side, so we'd know where it was going. Sal would scoop up a couple of bundles, and we'd take off. We were probably delivering about three and a half million dollars a month, cash money, from what was being brought in from overseas. And that was only being split between nine, maybe ten guys. Hy Larner was making phenomenal money for Chicago.

⊕ ⊕ ⊕

The first time I went to Panama I was pretty excited. After all, it was the first time I'd ever been outside the United States. Not that you would've known you'd crossed a border into another country when you were with Hy Larner. We didn't need a visa or passport, we never went through any type of customs or security checkpoint. Nothing whatsoever. We flew back and forth between Panama City and Miami like there was no such thing as a border. On the way down in the plane, the conversation I'd had with Mugsy Tortorella years before came back to me. That was when I'd started to understand what Hy Larner and Sam Giancana were about. But now, ten years later, I still wasn't sure I actually had a very good grip on it. For all the talk about the CIA and dope and guns, I hadn't seen any evidence of it. Not yet anyway.

I don't know what I expected Panama City to be like, but I can tell you this much—what I saw wasn't it. I guess I figured it would be more like Florida, more Americanized. I don't think I ever got over how poor most of the people were. In Panama, there were only two classes of people, the dirt poor and the filthy rich. The military guys had money, the Panamanian Defense Force guys had money, and of course, the people who called themselves aristocrats had money. But the average guy was barely getting by.

Gambling might be a long shot, but for those poor people in Panama, it was their only shot at breaking out of a backstreet hellhole, which I guess was what made all the slot machines we had set up at all the gas stations such a hit. Here were these people, half starved, with barely a shirt on their backs, and they're lined up in front of a gas station, waiting to play on one of Hy Larner's slot machines.

They were getting rolls of quarters, and one after another they'd step up and dump them in those machines. *Bing, bing, bing.* End of game. Next. And it would go on like that all day. And all night. Naturally, it was fixed. All of Hy's overseas gambling operations were tremendously successful, but in poor countries like Panama, he hit the mother lode.

⊕ ⊕ ⊕

One thing about Hy Larner, which is true for any Outfit leader, he was not a risk taker. Whenever he went with me and Sal to Panama, we always flew on one of his jets. Hy hated props, and he was scared to death of the water. He didn't want anybody knowing where he was in the sky, so there was never any flight plan. And you always had two pilots when he was on board; in case one of them had a heart attack, the other one could fly the plane.

If Hy wasn't with us, Sal and I would get one of the cargo planes out of Miami, load it up with machines and parts, and fly down to Panama. We'd come back with a cargo hold full of U.S. currency. Usually when we landed in Panama there would be a caravan of three or four guys in vans and cargo trucks waiting to unload the plane. There was never any checkpoint or customs. And it was the same in the States.

Once we landed in Panama, we'd head over to one of Hy's companies, either Juliano or Deverlandia, to see Victor Vitta, the main man. Like Hy, Vitta carried a diplomatic passport—which was a hell of an accomplishment for the son of a barber from the North Side of Chicago. Vitta was not only handsome, he was also extremely well educated and spoke fluent Spanish, Italian, and Farsi. He had a great sense of humor, too—which he hardly ever demonstrated in front of Hy or the Bastones. When they were around, Victor Vitta was all business.

Besides running all the machines and gambling in Panama, Vitta also ran the Hotel Bambito, a very exclusive resort casino, owned (behind the scenes, of course) by Hy and Manuel Noriega. They had the Miss Universe contest there about once every four years. The place was magnificent, with marble everywhere, tropical gardens, swimming pools, riding stables, restaurants, boutiques, not to mention a huge casino. Naturally, we were treated like royalty.

With a setup like Hy had going down there, it wasn't exactly a hardship to travel back and forth to Panama. As a member of Hy's operation, I went to Panama at least a dozen times between 1982 and 1987. And in those five years, tons of gambling paraphernalia—millions of dollars worth—were flown out of the United States. At the same time, hundreds of millions of dollars—the skim from Hy Larner's overseas gambling empire—were being flown back in.

With the U.S. president's "War on Drugs" going strong during this period, as well as the government's supposed crackdown on smuggling along the nation's borders, there seems no reasonable explanation for organized crime's amazing ability to continue to conduct business internationally—and so flagrantly at that. The team of Sam Giancana and Hy Larner might have managed to get in bed with several foreign governments and their officials in years past, but by the eighties there were crackdowns on that type of activity; times had supposedly changed. I will always believe that the only logical explanation for Hy Larner's uncanny knack for flying contraband in and out of the United States with no interference whatsoever from U.S. authorities wasn't that he'd outsmarted the government or paid some customs official more than the next guy— I believe Hy Larner *was a part* of the U.S. government. Or at the very least, working *with* the government in some capacity—much like Manuel Noriega was doing at that time.

Of course, I didn't come to that conclusion overnight. At first I didn't see any evidence that Hy was collaborating with the U.S. government or its intelligence agencies. And as for moving weapons or dope through Panama, forget about it. All I saw in the beginning was a bunch of slot machines and parts. Which is what I thought we were transporting when Sal and I were in Miami in 1986, overseeing a cargo shipment to Panama.

By the time we arrived at the hangar area, the plane was loaded and ready to go, but then we were told there were mechanical problems and the plane wasn't going anywhere. When Sal called Hy down in Panama and said there was a problem, Hy raised hell. He told Sal that shipment had to go out to Panama that day, one way or another. And he didn't want to hear any excuses, either, like it was Sal's fault some part fell off that plane. Hy was practically coming through that phone. Finally things seemed to calm down, and Sal hung up the phone. He said Hy was calling a friend of his to get things handled, but he didn't say who that friend was.

A few hours later, an executive jet landed, the type that could hold maybe eight passengers and a nice-sized cargo. But this wasn't your typical executive's plane. Olive drab in color, the jet bore the markings of the Israeli military. When four soldiers, dressed in uniform, stepped off and started transferring everything from Hy's cargo plane to their jet, I knew something was up. After they took off for Panama, I asked Sal what kind of friend had an Israeli military jet, and he just laughed and said, "A fuckin' Israeli general, that's who." Later I heard the general's name was Amiran Nir, and like Hy, he was tight with Noriega's associate Michael Harari.

Harari, a former Mossad terrorist and assassin, had a long history in Panama. From what I heard through Sal and Hy, Harari had originally gotten close to Omar Torrijos thanks to the Panamanian's Jewish father-in-law. Later Harari helped Noriega turn the country's National Guard into a sophisticated death squad, which—supposedly at Harari's suggestion—Noriega renamed the Panamanian Defense Forces.

Little by little, all the things I'd heard in the past about relations between Israel and Panama started to make sense. A few months after the general's plane had shown up in Miami, Sal and I flew down to Panama with another cargo plane. Once we got there, it was the same drill as always: the caravan of vans and trucks was there waiting for us when we landed. But on this particular day, Victor Vitta was late picking us up, so we had to hang around while the workers unloaded the cargo. That's when I saw the boxes of ammo. And the boxes of guns. It was just box after box after box. I felt Sal's eyes on me. He was watching me watch all those boxes come off that plane.

I finally understood what Hy Larner was up to. I might not know all the details, but I knew what was going down. Yeah, me and Sal had just flown a ton of weapons into Panama. Twenty-four hours later we were back in Chicago. Nothing was ever said about that shipment.

⊕ ⊕ ⊕

During the time I worked in Panama, I came to realize that Hy Larner's relationship with Manuel Noriega was a lot more convoluted than the typical underworld gambling partnership. I saw many more questionable shipments pass across the borders. I met Hy's Israeli friend General Amiran Nir on several occasions at the executive air hangar in Miami and at a number of social events. I saw Hy Larner travel frequently from Miami to Israel, often in Israeli military jets, often with Nir and Harari. At one point I was offered the opportunity to serve as a bodyguard for Hy whenever he went over to Israel.

Having been in Panama during the eighties, I will always believe that Hy Larner and the Chicago Outfit were tied to the Iran-Contra Affair. And I will also always believe that Hy was involved in the deaths of many individuals who either got in the way of the operation's success or threatened to expose it—one of them being Hy's personal Panamanian pilot, a man named Topia, who disappeared without a trace. Another was my old friend, the Chicago hustler Mugsy Tortorella.

Since leaving Chicago, Mugsy had been trafficking dope and, from what I heard, making a fortune. But in the mid-eighties, Mugsy got pinched by the DEA and shortly thereafter was seen snooping around the airport with several DEA agents. When the DEA dropped in on Hy Larner and started asking him a bunch of questions about Panama, I imagine that was it for Mugsy. Not long after, his body was found in a Miami warehouse. Mugsy Tortorella had been beaten and tortured to death.

CHAPTER 30

In the Outfit, when you screwed up, you got planted. End of story. It wasn't like they handed you a pink slip and you went to work for another crew. You were done. That is, unless you used a tactic that was a favorite with America's corporate set, the old CYA routine—cover your ass and blame whatever went wrong on the other guy.

That was Joey Lombardo's modus operandi, and that's what kept the son of a bitch alive and in power. Every time somebody who reported to Joey got in trouble, he just blamed that particular guy for the problem and got permission to have him whacked. Obviously, you didn't want to work for Lombardo.

The FBI's Operation Pendorf, initiated in 1979, was a great example of the truth of that statement. Pendorf stood for Penetration of Dorfman and turned out to be one of the FBI's most successful operations, demonstrating a relationship between Allen Dorfman (the mobbed-up consultant to the Teamsters Union pension fund) and organized crime. Over the years, a number of pension fund loans (ranging from two million to two hundred million dollars) had been made to casinos in Las Vegas. The FBI wanted to prove that Dorfman, as a behind-the-scenes representative of the

Outfit, owned a hidden interest (along with the mob) in these properties.

Thanks to the overturn of former President Johnson's executive order banning listening devices, in 1979 the agents installed wires in a number of locations around the country, including Allen Dorfman's office in Chicago. After two years, in May of 1981, the FBI had collected enough evidence to convince a grand jury to hand down indictments against five defendants in a bribery case, among them Teamsters president Roy Williams, Joey Lombardo, and Allen Dorfman. Found guilty scarcely six months later, they were facing some major time when New Year's 1982 came around.

It seemed logical to me that somebody was going pay for this Operation Pendorf screwup. And knowing Joey Lombardo, I had a good idea it wasn't going to be him—even though the pension fund and its officials were his responsibility. I'd heard that Jimmy Hoffa had been thrown in a vat of zinc in a Detroit fender factory, so I figured Williams and Dorfman might want to avoid assembly lines for a very long time. Of course, with the amount of time Dorfman was facing, there wasn't much chance he was going anywhere—unless he decided to turn. And that was the concern about Dorfman.

Around the end of the year, sometime in December of 1982, I went out to Las Vegas to make a pickup for Sal. That's when I first got wind that Allen Dorfman had a very serious problem. After I got back to Chicago, I started hearing the type of rumors that always circulated whenever a guy was going down. It was like there had to be a PR campaign to get everybody in agreement to take a guy out. I figured it was Joey Lombardo beating the drum.

Allen Dorfman was in trouble. Even Hy Larner, who'd been close with Allen's father, Red, back in the old days, was expressing concern that "the Dorfman kid" was too soft to do the time. But in January 1983 I had other things on my mind besides Joey Lombardo's damage control plans. We'd flown five Panamanian generals and their wives into town for a little two-day shopping spree—all on Hy Larner, of course. And it was my job, since I had a security company, to make sure the Panamanians were comfortable and safe.

I accompanied them on their shopping trips around the city. Hy had set up these enormous credit lines at different stores—jewelers, furriers, all high-end joints—and all I had to do was get the Panamanians there safely and then turn them loose. After they had their field day, I escorted them back to the hotel, which Hy wanted to be first class as well as out of the way. On this particular visit, we'd put the Panamanians up at the Lincolnwood Hyatt.

It was a very long day just standing around watching those generals and their wives spend Hy's money. By the time I left them at the Hyatt that first night, I was beat. It was after midnight and I'd barely gotten home when the phone rang. It was a very concerned Sal. *Alarmed* is probably a better word—like he was shitting in his pants. He told me to get over to his place right away and pick him up. We had to get to the Hyatt and move those Panamanians out of there. We couldn't wait until morning, he said. It had to be that night.

Sal and I beat it over to the Hyatt. I don't think he said a word the whole way. He was very tense. Once we got to the hotel, Sal turned into a storm trooper. He went from door to door waking up the Panamanians, telling them to hurry and pack up, that we were moving them to another hotel. Of course, none of them spoke a word of English. They didn't know what the hell Sal was talking about. But he didn't care. He was going to pull those Panamanians right out of their beds if he had to.

It was total chaos. We had ten people standing in the hall in their underwear, all yelling and talking at once in Spanish. Nobody knew what was going on. The men were screaming. Some of their wives were crying. And there me and Sal were, dragging them in their robes down the hall and out the front door. We threw those Panamanians and their luggage in a cab and away they went. By four o'clock that morning we had them settled in their new rooms in Rosemont.

I still had no idea what the hell had happened. I was just as much in the dark as those Panamanians. Things had been so crazy I hadn't even had a chance to ask. But when Sal and I got back in the car, I said, "What the fuck was that about?" He wasn't talking. He just said he didn't really know for sure. He told me Hy had called from Florida and said there was going to be a problem and that we had to get those Panamanians out of the hotel that night. He swore that was all he knew.

I didn't have to wait very long before I found out what "problem" Hy had been worried about. Around noon the next day, Irv Weiner and Allen Dorfman were on their way to have lunch at the Lincolnwood Hyatt. They were walking across the hotel parking lot when two guys waving guns ran up behind them. Weiner said later that one of the men yelled, "This is a stickup." Weiner ducked. Dorfman didn't. Problem solved. Nobody had to worry about whether Allen Dorfman could handle a long prison sentence after that. He was doing time in eternity.

⊕ ⊕ ⊕

In the midst of so much violence in my world, something amazing happened that May. A miracle occurred—my son Joey was born. But it hadn't been all sweetness and light. I can say that because I was by Sherry's side, dressed in a mask and paper gown, through the entire ordeal.

Throughout Sherry's pregnancy, I'd been concerned. It was her first child, and at thirty-two she wasn't exactly what doctors called "a young mother," so I knew there were greater than usual risks for her and the baby. When it came time for delivery, Sherry had a real hard time. After twelve hours, they called in a specialist.

When he was finally born, Joey was the color of a dusky grape; he was barely alive. Thanks to the doctor's use of forceps, there was an enormous indentation in his head and one arm hung by his side, like it was broken. When I saw him, all blue and limp—well, it was lucky I didn't have my gun on me—I went for the doctor. The entire hospital was up for grabs. Before it was over, they called in the police. But when they started to arrest me, I managed to calm down. I knew I'd be no help at all to Joey behind bars. And there wasn't any question in my mind that he needed me.

Joey's birth was the most wonderful—and most terrible—experience I'd ever been through. Or so I thought at the time. But the worst was yet to come. Four days later, when Sherry and Joey came home from the hospital, I knew something was wrong. I just didn't know what. He was listless and he barely cried, not at all like other kids I'd been around.

Joey's three-month christening was a big celebration. Sal Bastone became his godfather. Everybody was there, including my mother, who seemed unusually curious about her new grandson. After the service, when she had me to herself outside the chapel, she took me aside. There were tears in her eyes.

"Why didn't you tell me?" she demanded.

"Tell you what?"

"About Joey . . . why didn't you tell me? And coming here like this and not knowing . . ." Her voice faltered. "It's not right, Michael, finding out that my little Joey has Down's syndrome like this, at his christening. Well, I can't tell you what a shock . . ."

I didn't hear another word. If my mother was shocked by Joey's condition, then I was dumbfounded. I'd known something was wrong, but every time I'd asked Sherry what the pediatrician had to say after one of his regular checkups, she'd just tell me things were fine. I'd had no idea he had Down's syndrome.

As I soon discovered, Sherry had known all along. The doctors had informed her of Joey's condition shortly after his delivery. When I demanded to know why I hadn't been told, she claimed she'd been afraid that once I learned the truth, I'd leave them.

From then on, things went downhill. That was it for me and Sherry. Joey was all I had left. Of course, it wasn't the first time I'd become a father. But I made up my mind that this time I was going to try to act like one. And now, knowing that Joey had Down's, I vowed to do everything within my power to love him and to help him.

⊕ ⊕ ⊕

Throughout 1983, my world was in turmoil. And it wasn't just things in my personal life, with Joey and Sherry, either. There was the Masters case to worry about, the Outfit's problems in Vegas, the FBI investigations like Pendorf, as well as organizational issues of leadership and power. There was change coming, you could feel it in the air, and that made everyone, especially me at that time, very uneasy.

It had started shortly after Allen Dorfman was murdered, with the passing of Meyer Lansky. For months there'd been speculation about who would succeed Lansky in Miami, which was considered his stronghold and the link to all the international operations. The papers were saying that with Meyer Lansky gone, there might be a mob war, that maybe Chicago would try to muscle its way into Florida. And just who was the press saying was most suited for the job of handling organized crime's operations in Miami? Tony Spilotro.

But there was no way that was going to happen. For one thing, since the murder of Allen Dorfman, Tony had been considered a prime suspect and was being watched around the clock. For another—in a move that took everybody in the Outfit by surprise—that past January he'd been indicted for his role in the twenty-one-year-old M&M murders. The authorities had scooped Tony off the streets of Vegas and hauled him back to Chicago, where he was to await trial from the comfort of a jail cell. Then in September, while he was still in jail, Tony was indicted in the murder of a DEA informant. Given those obstacles to power, it looked to me like Tony Spilotro would be waiting a long time before he'd be taking control of anything, much less Miami.

And contrary to what some of the press were saying, there wasn't going be any war or any Outfit guys putting the arm on some New Yorker, either. That's because Miami wasn't a territory to be given away or fought over—Miami was inherited. Which meant that, thanks to his relationship with Lansky, Miami now belonged to Hy Larner.

Throughout his life Hy had lived by the rule that in the world of organized crime, along with fame goes *mis*fortune—and that was never more evident than when fifteen well-known bosses from five cities were indicted by a federal grand jury that October of 1983.

Now everybody knew that Operation Pendorf had been a real coup for the FBI, mostly because its success had led to another case, called Strawman, in which indictments were handed down against most of the Kansas City mob for skimming and maintaining a hidden interest in the Tropicana in Las Vegas. All the defendants were convicted (with the exception of the KC boss, Nick Civella, who died during the trial). By the time Meyer Lansky passed away in 1983, Kansas City was pretty much wiped out.

Unfortunately for the Outfit, during Strawman, the FBI had uncovered tons of evidence connecting the Chicago bosses to the guys in Kansas City. When the FBI started calling this new case against Chicago's top bosses Strawman II, it was pretty clear they were on a roll.

Although Strawman II put a crimp in Chicago's top guys, particularly Joey Aiuppa, at first they figured they could beat the charges. But then when witnesses starting lining up against them, they began to get worried. Three of the FBI's key witnesses—Allen Glick, the guy who fronted four Vegas casinos for Chicago; Aladena "Jimmy the Weasel" Fratianno; and the former Teamsters president Roy Williams—had everyone seriously concerned. From what I understood, they knew enough to bury just about everybody who was anybody in the Outfit.

And it was no use trying to take them out, either. Even if Lombardo (who was serving time on Pendorf in Leavenworth) or Spilotro (who was in a Chicago jail awaiting trial for murder) had been available to do the job, it would have been an impossible task. The more critical witnesses, like Roy Williams, had been under heavy security for months, ever since the convictions had come down in Operation Pendorf. So they were virtually untouchable.

Ultimately it wouldn't be just Williams, Glick, and the Weasel the Outfit would have to worry about; by the time the trial got under way in 1985, there were guys flipping left and right. It was pretty clear that the Chicago Outfit was going to take a major hit. Sal told me Joey Aiuppa figured he was going away for sure if he didn't get some better representation. At seventy-seven, Joey Aiuppa was an old man, and he didn't want to die in prison. He was desperate to walk away from those charges and wanted to bring in

Pat Tuite, an attorney who'd represented mob cases in the past. But Sal said that Aiuppa had run into a wall with Tuite. Supposedly, the big-shot lawyer told Aiuppa that he'd need a million-dollar retainer before he'd even walk in the door.

It might seem crazy, playing hardball with an Outfit boss like that, but Tuite had his reasons; he was far from stupid. He knew that Outfit guys had a reputation for not paying their attorneys. They'd get off and then leave the lawyer holding the bag. If the guy made any noise about his bill, it was "take me to court," which, of course, no one ever had the balls to do.

So now Aiuppa and his pals had a dilemma. They didn't want to go on their kick, take their defense money out of their own pockets. So what did they do? They decided to go to Las Vegas—the now crime-free town—and let their skim pay Tuite.

That's when Hy Larner entered the picture. Of course, I didn't know anything about Tuite's demands at the time. All I knew that day in September 1985 was that Sal and I were supposed to pick Hy up at the airport. So when Hy got in the car and said we were going to Sal's place on the lake in Wisconsin, I just figured it was another one of their secret meetings.

To say I was surprised when two of the Outfit's most powerful guys walked into Sal's place would be a major understatement. I had to figure Tony Accardo and Joe Ferriola hadn't come all that way for a barbecue. Something was up. Something big. The Strawman cases had put a lot of heat on everybody, but it was Accardo and Ferriola—who'd managed to escape indictment—who were really under the light. For months they'd been watched by the FBI, the IRS, and the DEA, so to meet for a sit-down in Chicago was out of the question. But Sal's place near Lake Geneva was perfect. It had an enormous garage, so visitors could pull in and close the door. Above the garage was a private meeting room with all the comforts of home. And to make everyone feel even more secure, there was a cop in town who was also Sal's buddy; any activity from law enforcement, and Sal got a call.

After everyone got comfortable, things got very formal. Accardo said they wanted me to fly out to Las Vegas, pick up a package, and bring it back. He said they'd call me when it was time to leave. With that, our meeting was over. No questions. No answers. No more discussion. Everybody just got up and left.

To say I wasn't exactly overwhelmed by the task would be an understatement. I'd been doing this sort of thing for years, so it wasn't a big deal to me. But these guys had made it a big deal, which led me to believe this wasn't the typical package.

I drove Sal and Hy back to Chicago, and the whole way nobody mentioned another word about my upcoming trip to Vegas. The three of us reminisced about Hy's anniversary party the year before at Chicago's Park Hyatt. We talked about the difference between the weather in Florida and Panama. Just social stuff. My curiosity was killing me.

A week went by, and I was getting nutty wondering what was going on. Finally I got a phone call from Sal, who was very mysterious. He said, "Meet me at O'Hare. Bring some clothes. We're gonna be gone a few days. . . . And by the way, don't forget your gun." Now the minute Sal said something about my gun, I knew I was on a mission. He'd never told me to bring a weapon along before. He knew I could carry one on a plane because of my job as deputy sheriff, but it had never come up until now.

When I got to the airport, Sal was there waiting at the ticket counter. And so was Hy, which really surprised me, because Hy Larner was no bagman. For him to personally fly out west to pick up a package for the Outfit was ridiculous—unless that package was the most important one that had ever crossed the continental United States. And as for going to Las Vegas with me and Sal, that had never happened.

As it turned out, we didn't fly to Las Vegas. We went to Salt Lake City instead and got a hotel about four blocks from the Mormon Tabernacle. It was a real classy place and had these great suites. So far it was looking more like a vacation than some top-secret Outfit deal.

We went down to dinner, and still no word on what was coming down. Then, out of nowhere, I see Joey Hansen—Tony Spilotro's man in California—standing in the lobby. Hy got up and went over and they talked for a while. Then Hansen left and Hy came back to the table and said we had to leave.

By that time it was after ten o'clock at night. We got a cab, and Hy gave the driver an address that, judging from the look on his face, was not in the nicest area. We ended up at this little run-down sleazy motel in the seediest part of Salt Lake City. I didn't like the looks of things at all. We got out and Hy told the cabbie to wait.

It was pitch-black out. I had my weapon ready to go. But still nobody had told me a thing. We walked up to one of the motel rooms and Hy knocked on the door. The curtains were pulled, and you could tell it was dark inside the room. I was very uncomfortable with the entire setup. Suddenly the door flew open and we had two sons of bitches with sawed-off shotguns standing right in front of us. Face-to-face. I didn't even take the time to see who it was. I had my shit. I was ready to whack them.

When I realized it was two of our own guys, Joey Hansen and Fat Herbie Blitzstein, I was tremendously relieved. Fat Herbie and I stood guard while Joey hauled two duffel bags to the cab. We threw them into the trunk and took off, leaving Herbie and Joey behind. The entire time, that cabbie was looking the other way. He didn't want to know a thing.

Back at the hotel, we carried the bags to Hy's suite. The minute we got them in the room, Hy said, "Put a chair against the door." Once we did that, he opened the duffel bags and dumped everything out on the dining room table. I could not believe my eyes. It was all cash. A million dollars in hundred-dollar bills, bundled into ten-thousand-dollar packages. There were casino wrappers around all the packages. And not just one casino, either—there were nine or ten different casinos' bands: Caesar's, Valley, Frontier, Freemont, Stardust, Desert Inn, Tropicana. We put on rubber gloves and got to work. We didn't count it, we just disposed of the casino wrappers and replaced them with rubber bands.

It was around this time that I discovered why we hadn't gone to Las Vegas as originally planned. Evidently Hy had intended to use his personal aircraft for this trip, but at the last minute there'd been mechanical problems. Hy's son Bruce lived in Salt Lake and was a commercial pilot. As Hy laid out his plan, I got the impression this may not have been the first time he'd utilized his son's position with an airline. Hy said I was supposed to switch bags with his son before we went through the screeners. I would be carrying a couple of bags that were identical to the ones Hy's son took on board when he was flying. Since there wasn't any screening for pilots, Bruce would just show his pass like he always did. We'd meet by the gate later, and he'd give me back the bags with the money. Once I had the money and Bruce was on his way, we'd be home free.

That morning we packed a million dollars in the two flight bags and headed for the airport. Everything went like clockwork. Bruce and I made the first switch, and I went through the screening. Then we met at the gate, and he gave me the bags with the money and I gave him the empties. And that was that. I had to hand it to Hy Larner; that deal with his son went smooth as silk.

Once I found my seat on the plane for Chicago, I realized that Hy had me sitting away from him and Sal. The entire flight, it was like they'd never seen me before. But when we landed at O'Hare, they were my shadow—until we got outside the terminal, where three rental cars, with drivers, were waiting for us. I took the cash and got in the middle car. Hy got in the front car, and Sal took the back. The convoy went straight to Bucky Ortenzi's house, where we dropped off the bags. I understand Tuite got his retainer later that night.

After Tuite was on the case, all the guys were sort of semijubilant. Everybody figured Tuite had it all handled. To Aiuppa and his codefendants, it was like it was a done deal, like they were all going to be acquitted. So you can imagine their reaction when they were all found guilty the following January—1986. I understand they were all sitting around their hotel room in Kansas City, ready to open a bottle of champagne, when the feds showed up to arrest them. And what about Tuite? What kind of explanation could he possibly have given for this result? I can't think of one that would've satisfied me—not after advancing him a million bucks for his legal fees. And I guess that's why, for the life of me, I've never understood why Pat Tuite didn't get whacked. Go figure.

⊕ ⊕ ⊕

If there was anything that had really threatened to bring down the Outfit over the past few years, it was a lack of strong leadership and control. If anything, putting Joey Aiuppa behind bars probably saved Chicago from a total crash and burn. And that's where things were headed, too, when Joe Ferriola stepped in as boss in January 1986.

If anybody could make things right, I figured it was Joe Ferriola. I'd known Joe for years, and he was a hell of a businessman. With him, you didn't move up the ladder by whacking people, you did it by making the green. Aiuppa couldn't see that. He would've let things continue going down. His deal was muscle. With his philosophy, you didn't make money, you got somebody on the arm and squeezed it out of them. The guys around Aiuppa always had their own agendas. They were into porn, prostitution, shakedowns—all very sleazy stuff. They didn't give a rat's ass about the machine business or the gambling business. They didn't have the slightest idea what kind of dough the Bastones' machines were bringing in. Their attitude was "Why do I want a bag full of quarters from some fucking machine?" Machines weren't macho enough. A lot of those guys were nothing but ignorant thugs with hair-trigger tempers. They might have been Aiuppa's boys, but with Joe Ferriola as boss, their days were numbered.

Joe had no patience with the bullshit that had been going on. Things had been falling apart in Las Vegas ever since Tony Spilotro hit the Strip. Before Tony went out west, Chicago had been bringing the skim home by the suitcase. With Tony it had gone from suitcases to briefcases to where it was now—in envelopes. Something was wrong. And Joe Ferriola knew exactly what it was—or should I say, *who* it was.

Joe had a guy in the Vegas police department who had been keeping tabs on Tony for a while. He got the word back to Joe about all of Tony's extracurricular activities, so Joe was up on everything going on in Vegas: the fact that Tony was screwing Lefty's wife, the little side deals he had going with burglaries and home invasions, the millions in hot jewelry, the juice operations, book joints, and sports betting deals, his involvement in the cocaine trade, and the one hundred plus murders.

Last but not least, there was Tony's handling of the skim out of Vegas. Evidently Ferriola had a tip that Tony might be stashing a little extra on the side, so he sent some guys out west to check it out. They burglarized Tony's house and found ten million dollars stuffed in the waterbed, money that should've been sent back to Chicago. That alone was reason enough to have the guy taken out.

On June 23, 1986, the brutally beaten bodies of Tony Spilotro and his brother, Michael, were discovered in an Indiana cornfield. The Chicago Outfit had finally cut its losses.

CHAPTER 31

I n the twelve months following Tony Spilotro's death, things were on the uptick. For most of the crews, business was booming. My old life as a cop was behind me now, and my new one—as one of the Bastones—was really starting to take off.

But like they say—what goes up must come down. I guess that's been the story of my life. It was May 1987 when I got a call from my lawyer, James Marcus. As it happened, I was sitting by a pool in Florida, sipping on a piña colada, when Marcus gave me the bad news: I'd been indicted by a federal grand jury on three counts of racketeering—racketeering conspiracy, racketeering, and extortion conspiracy—all stemming from my tenure as police chief in Willow Springs. Even worse, the charges fell under the federal RICO Act, so all together they carried some very heavy time—a maximum sentence totaling seventy years.

The RICO Act (the Racketeer-Influenced and Corrupt Organizations Act) was passed by the U.S. Congress in 1970, its main intent being to put the squeeze on organized crime. Over time, its application has broadened to enable a claim to be filed in federal or state court against just about anyone engaged in bribery and extortion. To get a conviction, the prosecution must demonstrate a pattern (usually several instances) of intent to defraud. As every wiseguy knows, there're a million scams that could fall into that category.

The other side of the RICO Act is its sting. If you get nailed under the RICO statutes, you'll not only go to federal prison for a very long time, but it's also possible you'll get hit for some major bucks—triple damages, plus legal fees—as restitution to the injured party or parties. That alone could take you broke—*if* you're found guilty. And that's exactly where I was sitting as I headed back to Chicago—with a great big *if* hanging over my head.

There'd been plenty of times when I'd known the feds were after me for one thing or another, but in this instance I never even saw it coming. I was still in shock when I got back to Chicago and went to the federal building for my arraignment. There was a huge crowd of media people with microphones, cameras, and lights set up outside the building, all waiting to see former Willow Springs chief of police Michael Corbitt. You would've thought I was the Boston Strangler the way they acted. With all the hoopla, it was looking more and more like a setup to me, like someone was planning to make a name for himself at my expense—namely, the prosecutor assigned to the case, Assistant U.S. Attorney Tom Scorza.

From the minute Tom Scorza and I laid eyes on each other, it was no good. It was personal. Scorza was a little Napoleon, balding with a beard, the kind of guy who doesn't have any hair on his head so he has to grow it somewhere else. He came across as very arrogant, puffing up his chest and swaggering around in front of the judge like "Look at me, I'm Mr. Big."

Right away I saw that Tom Scorza wanted me behind bars. He started by telling the judge, Prentice Marshall, that I was a flight risk and should be held without bond. But the judge wouldn't hear it. He said I had a family, that I'd been a police officer for twenty years, and then he set a personal recognizance bond at two hundred fifty thousand dollars. For that type of bond, all I had to do was go over to another building and sign a piece of paper. On the surface, that's all it amounted to—but as it turned out, it was a very unsettling experience.

I was immediately handcuffed and hauled over to the U.S. marshalls' lockup where they took my picture and fingerprints. *Then* I signed my bond. Although no one had proved a thing against me, I was already being treated like a criminal. I was feeling pretty sick about the deal, realizing that I was very close to becoming part of the system.

It wasn't twenty-four hours before I was approached by two FBI agents. We went to a diner, an out-of-the-way spot, and they proceeded to tell me that my life was in danger. While they must've known I was close with the Bastones, during our conversation they never mentioned Sal or any of the crew, or precisely who it was that had threatened my life. They just said the threat came from parties connected to organized crime and that they'd picked it up on a phone tap. They said things were looking pretty bad as far as my future health was concerned, but if I agreed to cooperate with Tom Scorza on the Dianne Masters murder case along with several other organized crime investigations, they'd guarantee my protection— and my family's—under WITSEC, the federal witness protection program. Naturally, I told them to forget about that. No way was I going to turn.

Thanks to that meeting with the FBI, I now knew two things I hadn't known before: one, I was definitely a suspect in the Masters case, and two, me and my family were in some major danger. Anytime you hear the FBI use the words *witness protection,* you can bet you've got a problem. It was like those agents had radar. As I'd seen my whole life, someone in my position could become a liability to the Outfit real fast. If you only stand to get a few years, it's no big deal. The Outfit's not going to worry, because everybody knows you can do that amount of time. But when the prosecution starts throwing around numbers like seventy years, which they had in my racketeering case, then everybody gets concerned.

Allen Dorfman was only the most recent example of how they handled that type of concern. The truth is, a guy facing major time might do anything. Who knows? And that's the question that makes everybody real nervous. The feds worry about you running, and the bad guys worry about you turning. I was out on a limb, a very long limb.

⊕ ⊕ ⊕

After I left the meeting with those agents, it was like everything changed. All of a sudden it felt like things were coming at me from all directions. Having watched Tom Scorza in action, I was starting to wonder if beating those charges was going to be as easy as I'd originally thought. After all, they'd gotten Jim Keating on extortion just the year before. He'd been convicted of accepting bribes in several vice cases. Maybe I was next.

Given what I'd just been told by the FBI, I also had to consider the possibility that somebody was going to come after me, and how that might hurt Joey and Sherry.

Since his birth, I'd made good on my vow to be a real father to Joey. When I wasn't working with Sal and the guys, he'd been right by my side. I had a beautiful Mercedes convertible, and I'd strap him in his car seat next to me and we'd take off. He thought he was a real big shot riding around with his daddy like that. He couldn't wipe the smile off his face when we'd pull into a Dairy Queen.

It was good to see my little guy smile like that. He was only three and he'd had a very rocky time. Along with Down's syndrome, he'd developed some very serious medical problems, including a dangerous heart condition. His doctors had also discovered that he was hard of hearing and had very poor eyesight. On top of that, he could barely walk. Or talk—at least so that somebody besides his mother and father could understand him. It broke my heart seeing Joey go through so much pain and difficulty. But he was a tough little guy. And through it all, he'd never lost that smile. But what about now?

When I got home, I immediately told Sherry I had to move out. It was a real sad deal. She was pretty shaken up. And as for Joey— he might be only three, but he understood that his daddy was in trouble, and I could tell that it scared him. As I started packing, he just sat on the floor looking on, big-eyed and whimpering, trying hard to be brave.

Seeing Joey like that was one of the low points in my life. I'd never thought things would come to this. But I knew that if I truly loved him, I had to go. It was the only way. Anything else wouldn't have been love, it would've been selfishness.

As I headed for the door, Sherry and I both tried to put on a brave face. But then Joey burst into tears and threw his arms around my legs and wouldn't let me go. With him all torn up like that, well, I couldn't help it—I started to cry, too. I knelt down and held him in my arms. More than anything, I wanted that moment to last forever. But I couldn't imagine doing anything that might jeopardize my little boy. I kissed him good-bye and walked out of the house. I know it might sound crazy, but for the first time in my life, I actually felt like a father.

⊕ ⊕ ⊕

Naturally, I didn't tell Sal about my conversation with the FBI. I told him I'd moved out of the house in order to keep Sherry and Joey out of the limelight, and he seemed to buy it. I acted like everything was fine. But I could tell that Sal was watching me closely now, to see how I was handling things with this new case against me. He knew I'd been under a lot of pressure in the past and that I'd handled it. There'd been my grief with the village when I was chief, the IRS, and Joe Testa's death and all the crap I'd gotten from Aiuppa about the money. Through it all, I'd managed to keep my head on straight. He knew that.

I had to believe that if Sal was nervous, it wasn't my racketeering case that had him concerned. Having all that prison time staring at me was certainly an issue, but I figured if anything had him worried, it was the potential spin-off from the case: the investigation that was going on into the canal and what might come of it, for starters. Although no indictments had been handed down, the authorities were still working the Dianne Masters murder. I hadn't told Sal that I now knew I was a suspect in the case, but he was well aware that it had the potential to bring down some major heat on everybody in the Outfit.

There was also the fact that I knew lots of details about lots of different Outfit operations. Sal had brought me into all sorts of things he probably shouldn't have: Hy Larner's international deals and the CIA, Las Vegas, Bally, the machine operations, Chicago's political corruption. Compared to that little book Dianne Masters had gotten ahold of, Michael Corbitt was a walking encyclopedia. Sal had to be worried about me. But the question was—worried enough to have me whacked?

⊕ ⊕ ⊕

After I moved out of the house, I became very cautious. I had my apartment and phone listed under an assumed name. I started driving different cars. I changed my schedule every day, watching for any sign I was being followed. I started carrying around two guns, sometimes three, so I'd be ready to take on anybody who made a move on me.

It was around this time that I started running into Gerry Scarpelli. The guy was everywhere, turning up like a bad penny. He was over by my apartment, cruising around the neighborhood. I'd see him downtown in a parking garage. One time I was driving along on the expressway and looked over at the car next to me and—whoa—there's Gerry Scarpelli. That's when I realized he was stalking me, that they were getting things prepared to set me up. Maybe it would be a big boom, like Joe Testa, or maybe they were going to kidnap and torture me and then kill me.

But even if that's what they were up to, I knew I had to keep up the front and continue working with Sal every day like nothing was wrong. If I gave even the slightest inkling that I was concerned, that would be it. Those guys were like junkyard dogs; one false move and they'd take me out.

⊕ ⊕ ⊕

Like Sal Bastone and his crew, Tom Scorza was watching me, waiting for me to trip up in some way. But unlike Sal, who I didn't believe had the heart to have me whacked, Scorza seemed hell-bent on bringing me down. And I was determined he wouldn't. Based on the prosecution's disclosures, I didn't think their case was strong enough to put me behind bars. On Count One against me, it was alleged that in 1982 I'd met with a small-time hoodlum named Larry Wright (who turned out to be undercover FBI agent Larry Damron) and conspired to take monthly bribes to allow an illegal bookmaking business to operate in Willow Springs. On Count Two, which was a restatement of Count One, it was alleged that Damron and his FBI partners were victims of extortion conspiracy. Count Three contained several elements, focusing on what the assistant U.S. attorney claimed was a pattern of criminal activities, in an attempt to show that my "arrangement" with Larry Damron wasn't an isolated incident. On this count, Scorza had all sorts of witnesses lined up to testify against me on everything from selling badges to extortion. There was even an instance where I'd supposedly aided and abetted in a tavern arson. In every one of these deals, all they had was some guy's word against mine. Talk about a bunch of ghosts from your past coming back to haunt you. And as far as I was concerned, you could see right through every one of them. Tom Scorza might've been determined as hell, but I didn't think he had a leg to stand on. He had no solid proof of any wrongdoing on my part. It was all hearsay.

On top of that, my so-called meeting with Agent Damron was never taped, which meant they'd have to use his field notes in court. Of course, Damron *was* FBI, and to a jury an agent's word is golden, so I was pretty sure that was where Scorza figured he really had me. But I wasn't the least bit worried about that meeting with Larry Damron: I had an ironclad alibi.

Early on in my case, the prosecution had tried to cut a deal with me, but I was confident I was going to beat the charges, so I wasn't at all interested. Like I said, I had an ironclad alibi: I could demonstrate that I'd been in Florida on the day Agent Damron claimed he'd met with me. And I had a rental car receipt to prove it.

I was certain things were going my way. But then, at my pretrial hearing, the prosecution presented evidence to the judge proving that my wife had doctored the date on the rental car receipt. Not that she meant to hurt me. She didn't. She thought she was helping.

Of course, when it came out that the evidence was a fraud—well, the judge was not amused.

And suddenly the case wasn't just about me; Sherry was involved, too. The judge called a recess, and behind closed doors, my lawyer informed me that unless I went along with the feds on the deal, Scorza was going to bring charges against Sherry. They were saying Joey would become an orphan. I couldn't let that happen. There was no other way. I pled guilty.

⊕ ⊕ ⊕

Tom Scorza may have gotten me in a corner. But he and I both knew that if the judge agreed to let me post bond, I'd be back on the street, a free man, until sentencing. And there was just no way he could let that happen. After I entered my guilty plea, Scorza proceeded to dirty me up even further. He told the judge that they'd uncovered enough information to convince them I was a flight risk. He brought out my financial statements and said that I had a fifty-five-foot yacht capable of leaving the country. When the judge didn't seem convinced, Scorza pulled out all the stops, saying that I was also under investigation in a major Chicago murder case.

Although Scorza hadn't come right out and said it, I knew what he meant, and so did Judge Marshall: I was being investigated in the murder of Dianne Masters. And with that little bombshell, Scorza had finally hit a nerve. The judge ruled that I was to be taken into custody and held at the MCC, the Metropolitan Correction Center, until sentencing—unless I could meet a two-million-dollar security bond, meaning I'd have to put up the entire nut, not a percentage.

By this time I'd realized that even if I raised that money, Scorza would get the judge to jack up my bond some more. I gave up; I decided to wait it out.

⊕ ⊕ ⊕

While I was in the MCC, I had a lot on my mind, especially Joey, who I hadn't seen for some time. I was also very concerned about Sherry managing money. She'd always relied on me; she didn't have the slightest idea about how to handle our affairs. Now, when she needed money, I had to have her pull ceiling tiles out of the house to find the places where I'd stashed it away.

Having Sherry in control of all that dough was a recipe for disaster. Gradually, much of what I'd made over the years began to disappear. She also dipped into my inheritance from Joe Testa, selling his condo for around half a million dollars. Where that money went, I'll never know. But as the weeks went by, I became concerned about our ability to meet our day-to-day obligations and my mounting legal expenses.

Looking back, James Marcus was not a good choice for me. I got him on the cheap because I figured I was going to beat the charges, a huge mistake, something I never should've done—not where my life was concerned. If I had it to do over, even if I had to beg, borrow, steal, and put my last dime on the table, I'd get the best lawyer I could afford.

When it came to sentencing, Marcus didn't put on any defense. It was only through the kindness of a bunch of folks from Willow Springs—and the understanding of the good-hearted Judge Marshall that I got as little time as I did. Marshall received almost eighty letters, from a variety of people in the Willow Springs community, on my behalf. I'll always believe he took those letters into consideration when Scorza asked for the maximum sentence of twenty years on each of the three counts. Marshall didn't go for it. He only gave me four years.

Believe it or not, by the time my sentence came down, four years sounded pretty good. At least it did until they hauled me back to the MCC and slammed that door on my ass. Then those four years might as well have been a million.

CHAPTER 32

I n the MCC, every day was an eternity. Of course, when you're waiting for the other shoe to drop, time goes real slow. And I knew Tom Scorza was coming after me for the Dianne Masters murder, which was further confirmed when the newspapers started saying I was a suspect in the case.

In the meantime, most of my family and friends—with the exception of Sherry and Joey and my mother—ran for the hills. It was especially difficult seeing my mother, a woman who'd always been tough as nails, so frightened for me. She'd always known a cop couldn't live the way I did on a few measly grand a year, but she'd never expected something like this, and she was heartbroken.

After I got to know how the system worked at the MCC—visitors checked in at around eight in the morning and had to wait for hours to see their incarcerated family member—I didn't even want my family to visit. And pretty soon, they didn't. And then I was completely, totally, one hundred percent alone. It wasn't a good feeling.

There was nothing I could do except sit and wait for Tom Scorza to make his next move. I was trapped on the segregation floor at the MCC for the next thirty months. At first I thought everybody went there, but then I found out that I was getting "special treatment" thanks to a direct order from the U.S. Attorney's Office—meaning Scorza. He said it was for my safety, but I knew there were about twenty cops in MCC's general population at the time. That may sound like a lot, but in a city with seventeen thousand on the force and a county with another seven thousand, it was probably pretty typical. It seemed obvious that by leaving me up on the sixth floor, all by myself, Tom Scorza wanted to turn up the heat and soften me up for when the Masters deal came down.

Being in a joint like the MCC takes away all your humanity, turning normal people into animals. And that includes the guards, who were real sadistic. Some of them were crazy, no doubt about it. They got a kick out of fucking with your head. Like they'd turn the lights off and then turn them on. Over and over. It didn't matter what time of day it was, either. They won't let you have a watch, so you don't know what time it is. Pretty soon you don't even know what day it is. You have no idea what's going on in the rest of the world.

The reality is, you're living in a cage. My cell was like something out of an old zoo that they don't allow animals to stay in anymore because it's inhumane. There was no heat. During the winter you'd need five or six blankets. It was so cold you didn't want to move. There was a toilet with a sink attached to it, but it never worked. There was no hot water. Anything you did was with cold water. When you got a shower, they'd take you down and lock you in this little box, all by yourself, and you might wait two or three hours for them to come back afterward. You'd get so tired of standing while you waited that eventually you'd end up sitting there, wet and naked, on that cold floor. No towel. Just cold walls and cold water all around you. Sometimes I'd get mad and start hollering. I could hear the guards laughing. They'd be just ten feet away, sitting in an office.

I was allowed one fifteen-minute call a week. If I was in the middle of a conversation and my time was up, too bad. They shut me right off. And mealtime was a real treat, too. The food came up, but I didn't get it right then. Hell no. I had to wait. The guard would set my tray on the floor, and before long there were cockroaches crawling all over it.

The guards did nothing but sit on the phone, bullshit with the female guards, hang out together, play cards, and harass the inmates. A lot of times they came in drunk. They'd get to the office and prop their feet up on the desk and go to sleep for four or five hours. If you gave them any problems—well, early on I got the word that they'd do you in a minute. They'd call their "response team," which was a group of guys wearing helmets and goggles and flak vests, and they'd beat the shit out of you—that's if you were lucky. They were totally brutal. I saw one poor guy who they got on real bad. He was a fighter. So what did they do? They put him in a special cell and shackled him to a steel bed. And not just for a few hours or a few days. We're talking weeks. Every day or so the guards came in and turned him over from his stomach to his back. He'd lay there crying while he pissed and crapped all over himself. It was really terrible.

⊕ ⊕ ⊕

I was finally allowed out of my cell an hour a day because they needed an orderly. I cleaned the floors and generally straightened things up. All in all, my situation seemed to be improving—as much as it could for anybody stuck in a place like the MCC. But then they got this bunch of Cubans in, and overnight the jail turned into an asylum. The Cubans weren't your typical criminals. These guys were mental. They had revolts, setting fire to the place every other day. They'd throw their own shit on the guards. Sometimes they'd all just beat on the floor or bang on the metal doors—for days and days and days. You'd wake up, and they'd still be doing it. Or they'd have the place filled up with smoke because one of them set himself on fire. And when they disciplined them, it just got worse. One day when I was doing my cleanup routine, I found one of those Cubans dead in his cell. He'd committed suicide. There he was, lying in a pool of blood. I called the guards. But did they give a shit? Hell no. They took their time.

There was one group of mob guys, organized criminals, that I'd never had much contact with until I hit the MCC. And that was the El Rukns. They were called the Blackstone Rangers in their early days, but changed their name to El Rukn—meaning "black stone" in Arabic—when their leader joined a Black Muslim sect. These motherfuckers were bad. We're talking some very heavy dope-dealing. A ton of murders. A whole network of activity all over the country.

There was a major investigation into their organization in Chicago, with about sixty members indicted for a bunch of stuff. Those guys were all going away forever. The U.S. Attorney's Office got around nine of them, including two of the top guys, to flip and go against their leader, Jeff Fort.

That's when they starting showing up on my floor at the MCC. They had the run of the floor. Their doors were open, they could use the phone whenever they wanted, they could have as many visits as they wanted, and it didn't matter how long, either—it could be maybe four or five hours. And believe it or not, they were having sex on the floor in there. The women were bringing in dope. The guys were always higher than a kite. When I was doing my cleanup routine on the floor, I'd see them doing lines of cocaine in the phone room. They were working drug deals from inside the joint. It looked to me like they could do just about anything they wanted. The guards saw all the bullshit. Everybody saw it. But nobody ever said a word.

I figured the only way that sort of crap could go on was by direct order of the U.S. Attorney's office. An attorney named Bill Hogan was in charge of the El Rukn case. And who did Hogan report to? My old pal Tom Scorza.

Knowing that, I wasn't real interested in running to the officials at the MCC and crying about all the special treatment the El Rukns were getting. I figured a guy in my situation might want to do the smart thing. I kept my eyes and ears open and my mouth shut.

⊕ ⊕ ⊕

Even if you try like hell to keep your nose clean, when you're in a joint like the MCC, eventually you're going to run into a problem.

In my case it was this black junkie who would not leave me alone. He was constantly harassing me about doing a heroin deal, telling me he knew this broad who could bring H into the MCC. He said he only needed five grand to buy it and wanted me to give him the dough. I figured he was nuts, so I blew him off. But that didn't stop him; he kept on me about it, and it was constant. I finally got tired of his crap and I told him to fuck off. But still he wouldn't leave me alone.

Things finally came to a head one day when I was on the phone talking to my kid Joey. Evidently one of the guards, thinking "Hey, won't this be funny," put that nutcase junkie in there with me. He locked the door and went on down the hall.

The junkie was like an enraged maniac. He just walked up and decked me. It was a hell of a shot, and when I tried to get up, he gave me another one. He was shouting, "Come on, mother-fucker . . . I'll kill you."

I could see through the window that the guard was back, but he didn't do anything but just stand there watching while that son of a bitch kept right on punching me. When I finally realized the guard wasn't going to lift a finger to help me, that did it—I went totally berserk, started punching and kicking, and finally got the guy pinned on the floor. Then I took that phone cord and wrapped it around his neck. I was wrapping it tighter and tighter. He was turning gray. He was dying. But I didn't care. Shit, I wanted him to die. I wanted to kill the son of a bitch. The next thing I knew, there were three guards on top of me. They dragged us both out and rushed the guy to the hospital.

I was still hot. I got up toe-to-toe with that guard who put him in there with me. I told him, "I wanna see your supervisor, mother-fucker. You set me up. I know it and you know it. So go get your supervisor. Do whatever you want to do. Charge me, I don't care. I'm ready for all of you bastards." I knew it was a setup. All those guards knew it, too. They took me back to my cell, and I never saw that one guard again. Apparently the guy I beat to hell managed to live, but he didn't come back to the sixth floor of the MCC.

After that little incident, everybody pretty much stayed away from me—which got pretty lonely. But that's really the only way to survive in a joint like that. It's kill or be killed.

⊕ ⊕ ⊕

Months went by and I had no word of what was going on from my attorney. One night somebody from another building in downtown Chicago fired a high-powered rifle through one of the windows. I laid there awake in my cell for a long time that night. My skin was crawling. Something snapped. All of a sudden I looked at those four walls and that was it. I had to get out of there.

From then on, everything changed. I started losing hope—which was exactly the frame of mind Tom Scorza wanted me in. That's how they play the game. They don't want you to be sharp and confident. They want you to be like putty in their hands. And then, after a while, they come around with their big smiling faces and suggest that if you cooperate, they'll help you.

I think I know extortion when I see it. And that was just how they operated. That's the primary reason the feds have such a high conviction rate. You either pled out or you went to trial, and then, if you were found guilty, you got the maximum sentence. After you spent a few months in the MCC, some guy in a suit comes along and says they'll cut that sentence you're facing by a third. You know they have witnesses who are going to testify against you. Unless you have the dream team, you conclude that it might be wise to plea-bargain and get it over with.

About that time, the papers were saying that indictments in the Masters case were about to come down. Supposedly, Scorza was coming after me, Jim Keating, and Alan Masters. Keating was already doing fifteen years on his other racketeering case, so when he showed up on the sixth floor at the MCC and I heard that Scorza was trying to get him to agree to testify against me and Alan Masters, I figured my back was going to be up against the wall. Shortly after Keating hit my floor, Marcus convinced me that if we went to trial on the Masters case, Keating was going to testify against me. Marcus recommended that I sit down with Scorza and see what we could work out.

I was such a fool. But like I said, I was also in that state of mind where I was pretty open to just about any way to get out of there. I ended up giving Scorza a proffer, which is a statement of fact that cannot be used against you at trial—although if you're found guilty it can be used at your sentencing. My proffer concerned the Masters case. I admitted to putting the car in the canal, but I said I did not know there was anyone in the trunk. I skirted the edges of some things. I didn't divulge too much.

As it turned out, Scorza may have been laying me and Keating off each other. Keating hadn't turned. But he also did a proffer, a vague one stating that he knew Masters and I were friends. He said he'd heard through other parties that we discussed something about disposing of Alan's wife. Really, it was nothing, but it did bring me into the scene a little more. And it did make it more of a conspiracy—which would become the most important factor in the case.

The only way I could've gotten clear of conspiracy charges was if I'd refused Alan Masters when he asked me to kill his wife and then gone straight to the police and spilled my guts. I would've had to bring up Charlie Nicosia. And the deal about the book. How Dianne was using it to blackmail Alan. How the Outfit wanted her dead—and Alan, too, if that book didn't turn up. Yeah, I would've had to give up everything. Had I done that, no trial. No case against me. I would've been out. Of course, I would've also been dead.

So I hadn't gone to the cops. But I didn't kill Dianne Masters. I dumped her car in the canal. And that made me an accessory to another part of the crime—illegally disposing of a body. In the eyes of the law, that's not exactly stealing an old lady's purse. It might not be murder, but we're talking a very serious offense.

The indictments finally came down in June of 1988. When they hauled me and Keating over to the federal building for our arraignment, I was almost relieved. By that time I just wanted to get it over with. I hadn't spoken to Jim Keating in years. Despite the bad blood, that day we started to talk. I think we both realized by that time just how intent Tom Scorza was on using the Masters case to make a name for himself. It was plain he was going to play it for all it was worth. Scorza wasn't just arrogant, he was also a grandstander. Keating and I had been around long enough to know that the press always loved that combination in a courtroom.

I left the arraignment knowing we were already being tried and convicted by the media. I immediately fired James Marcus and hired a new attorney, Dennis Berkson. Dennis turned out to be a total genius, a Perry Mason. He was terrific at strategy, too. He came over right away, and we started to prepare for the trial.

Once I got a look at what Scorza actually had on me, I was shocked. There was nothing. I couldn't believe they'd even managed to get an indictment. But right then I should've realized what Tom Scorza was about. I should've recognized him for what he was. After all, I'd known plenty of guys who wouldn't let anything get in the way of their ambition. It was just that they'd all been mobsters.

⊕ ⊕ ⊕

It was the first day of the Masters trial, and I hadn't seen Joey for months, but there he and his mother were—right up front in the courtroom. Joey looked real cute, too. He was dressed in a black suit, white shirt, and tie. When I was brought into the courtroom he gave me a kiss and a hug. It was all I could do not to cry. I'd wanted to see him, but not this way.

The second that Scorza realized Joey was in the courtroom, he made a motion to exclude him, claiming the only reason he was there was to elicit sympathy from the jury. I went ballistic. My attorney went ballistic. The entire courtroom went into an uproar. Judge Zagel was in a bad position. Finally he ruled that Joey could remain, but he had to sit in the back row. From back there, Joey couldn't see me, which was why he'd wanted to be there in the first place. He didn't come to court again after that. But I figured that was okay. I was certain I'd get off on the charges. And as for the time I had left on my previous conviction—I was eligible for parole in just nine months. So it wasn't like I'd never see Joey again. At least that's what I thought.

⊕ ⊕ ⊕

Although it became known as the Masters trial in the media, Alan Masters, Jim Keating, and I were never charged with the actual murder of Dianne Masters. Because there was no allowance for the charge of murder under federal law, the prosecution took a different tactic, charging us with racketeering conspiracy, specifically kickback bribery, extortion bribery, bookmaking bribery, and conspiracy to commit murder, which was where the murder of Dianne Masters actually came in.

Scorza intended to first convince the jury that the three of us had conspired over the years on a number of corrupt deals. Once convinced of that, it would be easy to persuade them that we'd also conspired on another matter—the murder of Alan Masters's wife.

From the very beginning, the trial was nothing but a total circus, a media dream come true. The case had everything—a beautiful woman; a rich, jealous husband; crooked cops; politicians; organized crime; and lots of sex. It also had a script. Only most people didn't see that. But I did. It was plain that Tom Scorza had one target: me. It wasn't Masters. It wasn't Keating. It was him against me, the man he called, with a sneer, "Big Chief Corbitt."

Once he'd built his case against us based on our history of racketeering conspiracy, or should I say, once he'd finished dirtying us up, Scorza began spouting his theory about Dianne Masters's murder.

Scorza contended that our original plan had been for me to drive Dianne's Cadillac to a crusher and dispose of the entire mess permanently, but that when I'd discovered the car had a flat tire, I'd had no choice but to dump it in the canal. He further theorized that with the change in plans, I'd realized that the car might be found and that Dianne's cause of death would most likely lead the authorities straight to Alan, so in an effort to draw suspicion away from him, I'd tried to make it look like she'd been abducted and shot to death.

There was also the insinuation that I might have tried to make her murder look like a mob hit, which was crazy as hell. After all, given Charlie Nicosia's involvement and the Outfit's extreme concern over Alan Masters's missing black book, only a fool would've played that card. Of course, Scorza didn't know that.

But that was his theory. And to my way of thinking, it was very weak. So were the prosecution's three main witnesses. None of them had any credibility, especially Robert Olson, who had actually been on my force. He swore he'd heard shots fired that night and then saw a guy who looked like me coming up from the canal. But every time Olson talked, he gave a different version of the events, so it wasn't hard to shoot holes in his testimony. Our attorneys ate him alive. His statements were all over the map. Olson was the only person who testified that he was an eyewitness that could put me near the scene. And once our guys got hold of him, that was the end of that.

The prosecution also tried to gain the jury's sympathy by painting this lily-white picture of Dianne Masters. Her charities and good works. How she loved her dogs, her daughter. They did everything they could to make her look like a saint. Of course, she wasn't. And to undermine the prosecution on that account, we brought witnesses in, including police officers, who saw her in the forest preserve parking with other guys while she was married. One witness got on the stand and testified he'd found her giving a guy a blow job in Alan's car.

After a few days of that type of testimony, you needed a ticket to get into the courtroom. Then our attorneys brought in the clincher: the tapes Alan had heard. In Dianne's own words, the jury heard the story of the wine being poured over her pussy and how her boyfriend had licked it off. At this point our attorneys put her boyfriend, Jim Koscielniak, on the stand. During cross-examination, one of our attorneys asked him what went better with sex, white wine or red wine? The whole courtroom went nuts—everybody was laughing. Except Scorza, who was screaming. It was total chaos. And the next day the headlines in the *Sun-Times* said something like ONLY RED WINE WITH SEX.

So now Scorza was desperate. He and his case were becoming a laughingstock. With no hard evidence in the Dianne Masters murder, he shifted the prosecution's focus back to the bribery and extortion aspects of the case. Now he began to hammer the jury with stories about our organized crime involvement, the chop shop business, bookmaking, kickbacks, and bribery.

But as we got to the closing arguments, I felt the only thing the prosecution had was Scorza's theories. And his grandstanding. Actually, if it hadn't been for the murder of Dianne Masters, Scorza probably would have backed off and never even gone to trial. His case was just too weak. By the time we got around to closing arguments, I really thought that, thanks to my new attorney, I was going to walk.

But not Alan Masters. I figured he was going for sure. He'd been free on bond throughout the trial, and every day he came in wearing a new suit, looking fresh as a daisy, while Keating and I looked like bums. For once, though, it seemed that appearances weren't going to matter. Scorza wanted the jury to look beyond those expensive suits and see Alan Masters as a very corrupt individual.

As far as James Keating was concerned, Scorza had brought out a lot of dirt on him, too. And therefore Keating was not handling the deal very well. He went down fast during the trial, and from what I could tell, the MCC had been particularly hard on him. Thanks to his previous extortion conviction, he'd already lost everything. He was completely busted, and his whole family had deserted him and moved out of state. His financial circumstances were so desperate that he'd had to depend on a freebie attorney arranged by Alan Masters.

As for me, Scorza had plenty of wrath—but little else. In my proffer, I'd openly admitted I drove the car and dumped it in the canal, but as I said earlier, a proffer can't be introduced as evidence in a trial, only at sentencing. Which meant that, aside from Robert Olson's testimony placing me at the scene, there wasn't one scrap of evidence against me. Nothing.

When the jury went into deliberations on Monday, June 12, I was feeling pretty good. But when they announced they had a verdict after only five hours of deliberation, I didn't know what to think. Everybody said that was a great sign, that we were in good shape. So I was still thinking I was going to be home free.

When the verdicts were announced, Alan Masters was convicted on two counts of racketeering, including conspiring to kill his wife, Dianne. Jim Keating was convicted on two counts, including the solicitation of murder for hire.

As for me, the jury found me guilty of only one of the original charges—accepting bribes in exchange for dumping Dianne Masters's Cadillac in the canal. It looked like I wasn't going to do any big time.

But when it came time for our sentencing about ten weeks later, it was obvious that someone forgot to tell that to Tom Scorza. In his lengthy and vicious attack on our character, Scorza said, "Judge, there is a big mystery in this case for me, and the mystery is how a just God allows people like Masters and Keating and Corbitt to crawl on the face of this earth."

Scorza portrayed me as a "prostitute with a badge," and he repeated his theory, in graphic detail, of what he insisted happened the night of Dianne Masters's murder—when I supposedly shot her in the head twice. He also introduced both Keating's and my proffers for the judge's consideration. Then he asked the judge for the maximum sentence on my conviction. He got it, too: twenty years, to run concurrent with my previous four-year sentence. Keating got twenty as well, to run concurrent with the fifteen he was serving. Alan Masters received the stiffest sentence of all: forty years and a fine of two hundred fifty thousand dollars. We were led out of the courtroom in handcuffs.

It was August 24, 1989. I was forty-five years old. Most guys that age would say they still had their whole life ahead of them. But mine was all behind me now. I'd left it, along with the people I loved, on the other side of a prison door.

CHAPTER 33

La Tuna is in the middle of the New Mexican desert, fifty miles from nowhere. In the summer it feels like one hundred and twenty degrees. In the winter, after the sun goes down, a man can literally freeze his nuts off. The sky is almost always blue in that part of country—no clouds, just a totally empty space above your head—and the desert sizzles out in front of you for miles until it disappears in a silver glare of no-man's-land where earth and sky come together.

Originally the Federal Correctional Institution at La Tuna was used as an internment camp for Japanese-Americans during World War II. After that, the Bureau of Immigration and Naturalization Services moved in. By the time I arrived on the scene in 1989, La Tuna had been converted to a medium-security federal prison.

There were virtually no black inmates—or whites, for that matter—at La Tuna. I was behind bars with what seemed like nine million Mexicans, all of them killers and gangbangers. If you're one of the few gringos, like I was, your only chance is to make everyone think you're a wiseguy. Otherwise they'll cut you in a second. It's not that they're scared of what an Outfit guy can do to them. It's that they're afraid of what your people will do to their families. So I let them think I was connected. But mostly I kept to myself.

After entering La Tuna, I didn't see any friends or family for a very long time. I kept in touch with my mother, whose health had been going downhill for some time, but with the exception of her and my sister Cecilia, everyone else in my immediate family had disappeared. Sherry had run through most of my money. There was nothing left between us except Joey. I called him religiously every week, but even with Joey things were different; I felt like I was becoming a stranger.

At first it hurt, having everybody run for the hills like that, but when the FBI agents informed me that they'd obtained hard evidence—a tape recording—of a threat against my life, I decided it was probably for the best. Of course, since the agents had refused to let me hear the recording, I couldn't be sure it actually existed. In the back of my mind I figured they could've made it all up, hoping to get me to flip and go witness protection. But I didn't know that for certain, and even the slightest possibility that the threat might exist made me sick with worry for my family, particularly my son.

I'd been incarcerated at La Tuna for two years before I saw Joey. By that time—1991—he was eight years old. Since I'd been away, he'd learned to talk and walk pretty much like other kids his age. There'd been birthday cakes and candles and Christmas trees and a ton of other stuff—happy, wonderful memories—that we'd never be able to share. There'd been some real tough times, too—surgeries and doctors and hospitals, long nights on white sheets when Joey had needed a father's hand to hold, and I hadn't been there for him.

When Sherry called to say she was bringing him down to La Tuna for a visit, I was real excited. I was also very nervous; I had no idea what to expect. On the day they were set to arrive, I couldn't stop pacing. And when the guard called out my name over the loudspeaker, saying I had visitors, my heart actually skipped a beat. By the time I walked into the visiting room, I was trying hard not to cry.

As I searched the crowd of inmates and visitors for Joey's face, I reminded myself that a kid his age grew fast in two years; Joey would be a different little guy than the one I'd left behind. Maybe I wouldn't recognize him. Maybe he wouldn't recognize me. Or maybe he wouldn't want to. But then a blur of arms and legs came flying through the air, and there he was, snuggled in my arms, planting those same slobbery kisses all over my face that I remembered from years ago.

During our three-hour visitation, Sherry sat on the sidelines, waiting, while Joey and I got reacquainted. Although prison regulations stated that inmates weren't allowed to touch their visitors and vice versa—it's one kiss hello and one kiss good-bye—if those guards had tried to enforce that rule with Joey, it would've been World War III. He hardly left my side the entire time. You couldn't pry that kid off of me. He sat on my lap. He hugged me. He kissed me. He sang to me. It was me and my shadow, just like it had been before I went away.

There's always a lot of children coming to prisons for visits with inmates, so the administration has plenty of stuff for kids, things like puzzles, games, and storybooks. Although Joey couldn't really read, he loved books, so when he saw *The Story of Chicken Little* on one of the shelves, he grabbed it and announced that it was story time.

I must have read that book aloud to him a dozen times that afternoon. Pretty soon we were chasing each other around the visitors' yard, and Joey was doing his best imitation of Chicken Little, giggling and yelling, "The sky is falling, the sky is falling." And later, when a guard announced that our time was up, that's exactly how I felt, too—like the sky was falling. Seeing Joey that day was one of the happiest times of my life. Seeing him leave was definitely one of the saddest.

After Joey left, I held on to the Chicken Little book. I thought of it as a souvenir of our time together that day. I kept it hidden under my pillow in my cell. It was a small thing, but seeing it every night always made me smile. Even more important, it reminded me that I had a reason to get up every morning—Joey.

⊕ ⊕ ⊕

Another year came and went at La Tuna. I talked regularly to Joey and my sister and mother during that time, but with every passing day, what family I had left seemed to grow more distant. It wasn't their fault; when you're in prison, even the people you're closest to eventually fade away. After all, the only thing you have in common is the past. And with each year behind bars, that past gets further and further away.

The truth was, my future was *inside* that prison. La Tuna was my world now. And it was a highly regimented one, too, with its own rules and rhythms, which is why—when two burly guards hustled me out of my cell one night in the fall of 1992 and said I was going on a "special trip"—right away I knew I had a problem. Nobody ever took a prisoner out at night; it was against all the rules. Something was up. And whatever it was, I didn't think I was going to like it.

After waiting for hours in the prison holding room, wondering what the hell was going on, I finally saw three U.S. marshals march in. As they handcuffed me to a belly chain and strapped a pair of leg irons around my ankles, they were all business—they wouldn't even look me in the eye. I asked them where they were taking me, but they didn't bother to answer. They just shoved me out the prison door and into an unmarked car.

When we took off down the empty highway doing about eighty miles an hour, my worst fears seemed to be confirmed. I figured they were taking me to the type of place where they could drop a guy. They'd dig a shallow hole, throw me in, end of story. I'd be just one more prison escapee who got lost in the desert. Again I asked the marshals where we were going. And again all I got was the silent treatment. Now I was pissed. I started yelling, demanding to know where they were taking me, calling them every name in the book. But that didn't faze those sons of bitches. It was like I was invisible.

Then, right out of the blue, the driver slammed on the brakes and made a sharp turn, and we lunged off the pavement onto a dirt road. Suddenly we were headed down a road that led to nowhere. We were really moving, too—going fifty, maybe sixty miles an hour. We scraped past cactus and boulders and barely missed the twenty-foot drops and arroyos that crisscrossed the open desert. When I saw headlights flashing off and on up ahead, I figured this was it for me. As we got closer, the glare of the headlights gave way and a battered pickup came into focus. We pulled up alongside it, and a man in a flak vest jumped in.

After we took off and I got a good look at our new passenger, I couldn't believe it. I knew the guy; he was part of the SIS from La Tuna. SIS is the Special Investigative Service of the Federal Bureau of Prisons. They don't get involved in normal prison operations, just serious criminal matters. I couldn't imagine why the SIS would be interested in me.

Pretty soon we turned back onto the pavement, and there was La Tuna looming ahead. When they took me through a side entrance and into a concrete building with towers on the roof manned by armed guards, I still didn't have a clue what was going on. And the more I thought about that—and what those U.S. marshals had just put me through—the madder I got. They took off my chains and escorted me up some stairs to a concrete room above the warden's office, where they informed me that I was now in protective custody, thanks to Assistant U.S. Attorney Tom Scorza. They said Scorza had informed the warden that I was entering the witness protection program. When I heard that, I went totally nuts. I told them there was no way I was in WITSEC. But no matter what I said, they wouldn't believe me—or maybe it just didn't matter. They escorted me to my "suite" and that was that.

I was held in solitary confinement at La Tuna for one hundred and twenty days. At first I was denied phone calls. Later, when my phone privileges were restored, I was instructed not to tell anyone, including my attorneys, where I was being held. During this time, FBI agents from the Chicago office came to see me. They were investigating the Bastone crew and Hy Larner and wanted me to talk. I told them that before I'd say a single word about the Outfit, they'd have to do two things—one, get me out of protective custody, and two, play the tape recording and prove to me that there really was a threat on my life. When they said no to both, I told them to fuck off.

Finally, four months after I'd taken that "special trip" into the desert, the warden found out that I really wasn't in the witness protection program. He was furious, too. I imagine his career flashed before his eyes when he realized that they'd violated at least half a dozen of my civil liberties by holding me like that. As far as the warden was concerned, I was a major hot potato. Overnight, I was on a plane headed for federal prison in Tallahassee, Florida.

⊕ ⊕ ⊕

My situation at La Tuna had been so isolated that in 1993, when I was thrown into the general prison population at Tallahassee, it almost felt like freedom. But of course, the Federal Correctional Institution at Tallahassee was just as big a hellhole as La Tuna, *hell* being the operative word.

If the New Mexican desert had been hot, Florida was a damned inferno. There was no air-conditioning in the prison, just some huge industrial fans blowing and rumbling so loud that you couldn't hear yourself think, twenty-four hours a day, across an open dorm that looked more like a chicken house than a shelter for human beings.

The dorm held two hundred inmates inside one hundred open concrete block "cubes" with walls five feet tall. Inside each cube were two bunks. There were no doors to the cubes and no toilets or sinks, just a steel mirror on one wall. The bathroom and shower facilities were at the end of the building, which was a hell of a long walk if you had to take a leak.

Unlike La Tuna, Tallahassee's prison population had very few Hispanics. It was mostly blacks from the Deep South and wiseguys from out east. Each dorm had one guard, one counselor, and one case manager, and not one of them was anxious to leave the safety of his office and mingle with a general population made up of cutthroat gangs and mobsters. So there wasn't any real security for the inmates—something I learned the hard way when three gangbangers walked into my cube one night and attacked my cellmate with a makeshift knife and a screwdriver.

They were holding my cellmate down, cutting him to ribbons, and blood was flying everywhere. The poor guy was screaming like crazy, but nobody was paying any attention, and it was becoming pretty obvious that the guard wasn't going to run down there and risk his life to save a lousy inmate.

I knew that if something didn't happen soon, they were going to do the poor guy right there in front of me. I jumped in. By the time the assholes from security came around, I'd beaten the hell out of those gangbangers. But I'd also gotten a screwdriver in the shoulder, and my cellmate had stopped breathing. The guards carted him off to the hospital, and that was the last we saw of him. I never heard if he lived or died.

As terrible as the fight in my cube had been, there was a positive side to the incident: it solidified my reputation with the other inmates as a tough guy, which is what you want if you're going to make it in a prison filled with killers, druggies, and gangbangers. After that, I managed to hook up with some of the East Coast wiseguys, among them a mob up-and-comer out of Nicky Scarfo's Philadelphia crew, who everybody called "Johnny Chang."

Under Johnny's supervision, the Italians had the prison all sewn up. It was unbelievable; they controlled just about everything but the locks on the doors. They were in charge of the telephones, the vending machines, the yard, and the visiting room. As long as they weren't too flashy about what they were doing, nobody screwed with them, including the officials and guards.

Once I got in tight with Johnny Chang, my situation at the prison improved dramatically. I got a job in the phone room, so I could talk to whoever I wanted, as long as I wanted. It was great; I called Joey every day, sometimes twice a day. And thanks to Johnny, my diet even took a turn for the better; believe it or not, the Italian inmates had a shed fixed up out in the yard where they could cook all their favorite Italian meals. Johnny had guys working in the prison kitchen who'd steal whatever he wanted and trade him for cigarettes or dope.

Being with the Italians at Tallahassee was like something out of *Goodfellas*. We had music, we had food—you name it, whatever we wanted, we had it. And if we didn't have it, Johnny Chang could get it. If it hadn't been for the fact that I was in a federal pen, it would've been a hell of a good time.

⊕ ⊕ ⊕

In the spring of 1993, just as I was settling in to life at Tallahassee, all hell broke loose in Chicago. The El Rukn trial had gotten under way, and there'd been a number of allegations that El Rukn inmates at the MCC had received illegal and inappropriate favors—including sex and drugs—from the U.S. Attorney's Office in exchange for their testimony against fellow gang members.

Right away the authorities launched an official misconduct investigation into the U.S. Attorney's Office, placing the chief prosecutor, William Hogan, and my old pal, his supervisor, Tom Scorza, on the hot seat. In connection with this investigation, everyone who'd been on the sixth floor of the MCC during the El Rukns' incarceration was formally interviewed. Everyone but me, that is.

I was told by an FBI agent that when the judge compared a list of the names of those interviewed with a list of the prisoners who'd been on the sixth floor during the time in question—and discovered I was the only witness who hadn't been formally interviewed—I'm told he hit the roof. Shortly after that, I was flown to Chicago.

I'd thought I was just going for an interview; instead I was thrown into a full-blown hearing. The attorneys drilled me for five hours on the stand about what had happened on the sixth floor of the MCC. In graphic detail, I testified that I'd seen the El Rukns using drugs and having sex with visitors. Of course, I wasn't the only person to have witnessed this outrage; other witnesses substantiated my testimony. And thanks to that, the government's case against the El Rukn gang went down the drain.

It was a very ugly scandal. The papers made me out to be the big man who'd buried crime-busters and rising stars Tom Scorza and Bill Hogan. I guess it shouldn't be a surprise to anybody that a few months later, when I heard that Scorza and Hogan had left their positions at the U.S. Attorney's Office, I didn't shed a single tear.

⊕ ⊕ ⊕

After I testified in the El Rukn case, FBI agents out of Chicago began coming to Tallahassee to "interview" me more often. I think they thought I was warming up to the idea of working with the authorities after my testimony in Chicago. Of course, they were wrong. But no matter what I said, they didn't seem a bit discouraged. They must've come down to Tallahassee a dozen times, trying to persuade me to go WITSEC and testify against my old Outfit pals.

The way the agents talked, catching a criminal in Chicago had been almost impossible those past few years, especially when it came to bringing Hy Larner and the Bastones to justice. One of the things they'd focused on in their investigation was Chicago's machine rackets, specifically video poker. But without more to go on—like having me as a key witness—the agents said they didn't have much of a case. At least that's the story they gave me.

Of course, I couldn't have cared less about the problems the FBI was having with an investigation back in Chicago. The way I figured it, those agents would just use me to make their case, then they'd get a big promotion and be on their way. Besides, what had they ever done for me? In all the time they'd been coming down, I'd only asked for one small favor: to hear that tape recording. But every time they'd turned me down.

After months of sitting through these interviews that never went anywhere, I was starting to wonder why the agents even bothered. It was always the same: They'd ask me some questions about organized crime in Chicago, and before I agreed to answer a single one, I'd say I wanted to hear the tape. When it was clear we were at an impasse, they'd bring up the witness protection program. I'd tell them they could forget that idea. And that was that, end of interview.

I have to admit those guys were persistent. Or maybe it was just that lack of success didn't matter to the FBI. After a while, they even started playing it like we were friends. But I never forgot what they were really up to. If they could convince me to spill my guts, it was one more step up the ladder for them—and I was that step. Friendship never entered the picture.

⊕ ⊕ ⊕

If the FBI agents had really wanted to get on my good side, they couldn't have played it more wrong than they did in the fall of 1995 when I learned that my mother was dying.

I hadn't seen Mom since I'd gone to prison six years before. Of course, we'd kept in touch through cards and by phone. So I'd known she was sick, but I hadn't realized just how sick—and I sure as hell hadn't expected her to die, not now, not while I was away in prison. But that's what my sister called to tell me in October. Our mother was losing ground, Cecilia said. And she was asking to see me one last time before she died.

I'd barely hung up from my sister's call before I was dialing my mother's hospital room. On the phone, Mom sounded real weak, not at all like her usual take-charge self. I told her I'd see her soon, that it was just a matter of filling out some paperwork and I'd be there. In no time at all, I said, I'd be on a plane to Chicago.

It wasn't a lie. Inmates went to see dying relatives all the time. True, there was an incredible amount of red tape a prisoner had to go through to get approval for a trip like that, but from what I'd seen at Tallahassee, nobody ever got turned down. They put you on a plane with some U.S. marshals and got you there, no questions asked.

But not me. Right away, my request was denied by the prison officials. When I called to give Mom the bad news, I could tell she was trying real hard not to cry. I tried to be optimistic, telling her not to worry, that I'd get things squared away. And then I went right back to the administration and put in another request. And again, without so much as an explanation, I was turned down.

For several months it went on like that. I'd call Mom and swear to her that I'd be there real soon, and then I'd go fill out some more papers and try again. By Thanksgiving it was obvious that no matter how hard I tried, or what I said, I wasn't getting the approval to go to Chicago to see my dying mother. And I knew time was running out.

I'd been telling Mom to hold on, that I'd be there soon, but I knew she wasn't going to be able to hang on much longer. I was talking to her every day, and she was really starting to go down. Finally, in desperation, I turned to my pals at the FBI and made them an offer: I'd cooperate with them on their investigation *if* they'd arrange for me to see my mother. But right away I could see that deal was going nowhere. It was just one excuse after another. They'd always talked a good game, but when the chips were down, it looked like they weren't going to lift a finger to help me. It was a slap in the face I wouldn't soon forget—or forgive.

On December 10, when I called Mom at the hospital, I could tell she'd taken a turn for the worse. Whenever we'd talked before, she'd listened quietly while I filled the time with a bunch of meaningless chitchat, which was my half-assed attempt to keep her spirits up. But this time was different. This time she did most of the talking.

"I sure wish I could've seen you, Mike," she began.

"Don't give up on me yet," I said, doing my best to sound cheerful. "I'm still working on it. You watch, I'll be flying up to Chicago any day now." Of course, we both knew that wasn't the case by that time, but I just couldn't bring myself to come right out and say it. Giving her a little hope was about the only thing I'd been able to do those past few weeks. I didn't want to take it away now.

Mom had always known the score; in my entire life I'd never been able to fool her. And I couldn't this time, either. She gave out a long, tired sigh. "There's no use in us pretending," she said. "Besides, this is long distance, we don't have the time or money to waste on tall tales. Sure, I'm disappointed we won't be seeing each other"—her firm voice went suddenly soft—"but I'm not disappointed in *you*, Michael. I know it's not your fault. And I don't want you feeling guilty about it after I'm gone—"

"Gone?" I said, cutting her off.

"Gone," she repeated. "When your time's up, it's up. And it looks like my time on this earth is just about done. You know it and I know it. No sense acting any different."

I felt the tears well up. "Listen, Mom, you just hang on a little bit longer." I was pleading now. "The minute we get off this phone, I'll go see the prison counselor. Maybe I can get a plane out of here tonight. You just hold on . . . I'll be there . . . you'll see."

"That's what I'm trying to tell you, Michael," she said, her voice cracking as she choked back a sob. "It's no use. I can't hold on any longer. It's time. . . ."

For some reason I don't remember hanging up the phone or saying good-bye to my mother that day. I guess I couldn't accept the finality of that word—not where my mother was concerned.

As tough as she'd been on me, I'd always respected Mom. Sure, she'd been critical of my lifestyle over the years, but I'd always known she loved me. She was a strong woman, with a good heart. And for the past few months, she'd been fighting off death, hoping to hold on just long enough to see me one more time.

When the call came in to the prison later that night, I knew her fight was over. A guard took me down to the counselor's office and handed me the phone. "It's Mom," Cecilia said as he closed the door, "She's gone."

⊕ ⊕ ⊕

Inmates were routinely released to attend a parent's funeral. But I wasn't going to take any chances. After I filled out the paperwork, I ponied up the entire cost of my trip to Chicago for Mom's funeral—including all expenses that would be incurred by the U.S. marshals who had to accompany me. And sure enough, this time I got the approval.

In less than twenty-four hours, I'd packed my bags and was sitting in the prison's transfer station. I got there early. I didn't want any slipups. I was going to be ready to go when those marshals arrived.

An hour went by and I asked the guard what time it was. After two hours, I was walking the floor. My scheduled departure time came and went. And still I waited. By that time I knew something had gone wrong. I just didn't want to believe it.

At last a guard shuffled in. The look on his face said it all. He hung his head. "I sure hate to be the one to break it to you, Mike," he said. "But that request to attend your mother's funeral . . . it was just denied."

I couldn't believe my ears. "Denied? How the hell can they do that? Shit, they can't—it was already approved. I even paid for the fuckin' trip, right out of my own goddamned pocket." I balled up my fist, ready for a fight. "Those motherfuckers are gonna pay for this bullshit."

The guard shook his head. "This didn't come from the warden's office. Some big shot back in the U.S. Attorney's Office in Chicago put the screws to the deal." He gave me a sympathetic smile. "I'm sorry, Mike. I'm real sorry."

⊕ ⊕ ⊕

After the agents sat down at the table in the prison conference room in January of 1996, the first thing they did was apologize about what had happened with my mother. They said their hands had been tied, that their superiors hadn't allowed them to intervene on my behalf. Then they opened a briefcase and put a tape recorder on the table. "We want to make it up to you, Mike," one of them said and hit PLAY. I held my breath. I'd waited seven years to hear this tape. Now the moment of truth had arrived.

I don't know what I expected, but I guess I'd never really bought all the crap those agents had fed me back in 1989 about catching an Outfit threat to me and my family on tape. Sure, ever since they'd first told me there was a problem, before I'd even gone to jail, I'd taken precautions to protect Joey and Sherry—just in case. But in my heart I'd never been able to accept the idea that my best friend, Sal Bastone, had betrayed me. And as for Sal being party to anything that might hurt Joey, well, that had been totally incomprehensible. Hell, Sal was Joey's godfather.

Now, after hearing the tape, I knew the truth. I'd heard it with my own ears: Sal Bastone had given Gerry Scarpelli the order to take me out. And he hadn't been concerned whether Joey got killed in the process, either. There was no way I could deny it—my "best friend" had not only betrayed me, he'd turned his back on my son as well. A thousand emotions swept over me. Scarpelli was just doing a job. But Sal Bastone? I wanted to kill him.

I tried to conceal my rage, but I was ready to explode. When the agents brought up WITSEC, my anger spilled over to them, and I knew I had to get the hell out of the conference room before I did something crazy. I stood up and headed for the door.

"Hold on, Mike," one of the agents said. "Come back, have a seat. What's the big surprise? You had to know those guys were going to come after you. Shit, you know Sal Bastone. You know what he's about. And you know Hy Larner, too. How many guys have they taken out? Why should you and your kid be any different? You want revenge? There's a million ways to get those sons of bitches. Cooperate with us on the deal, go into WITSEC. We'll take them down together."

What those agents didn't understand was that I'd never believed Sal would do that to *me*. Okay, so maybe I'd been in denial, but thanks to that tape, I wasn't anymore. All those years with Sal, with

Hy Larner—what was that about? What had Sam Giancana told me years ago? Don't forget who your friends are? Even guys like Sam fell victim to that delusion. When he turned his back on Tony Spilotro and got a few bullets to the head, that should've been proof enough that loyalty didn't exist in the Outfit.

But when I stormed out of the conference room that day, I didn't want proof of anything—all I wanted was revenge. And I wasn't at all convinced that working with the FBI was the way to get it. There was just one thing I was sure of: I was going to bring down Sal Bastone and that son of a bitch Hy Larner. Whatever it took, I was going to do it.

⊕ ⊕ ⊕

To achieve your goals, sometimes you have to do things you don't necessarily want to do. I might've wanted revenge, but eventually I also had to face facts; I was in prison, which meant my hands were tied when it came to personally going after anybody in the Chicago Outfit. The more I thought about it, putting Sal and Hy behind bars for the rest of their lives seemed like the only way I'd ever bring them down. I might not be able to kill them, but at least I'd have the satisfaction of knowing they were dying a little bit each day in a prison.

I told the agents I was ready to talk, and right away I threw myself into helping them go after Hy Larner's operations and the Bastone crew. That was all I could think about. I spent my days making phone calls, gathering information, and nailing down details about their business dealings. In the evening I sat in my bunk writing down everything I could think of that might be pertinent to an FBI investigation, giving special attention to my experiences with Larner and the Bastones. Pretty soon I had a box full of notes. And the other inmates were starting to get nosy. In 1997, after several threats on my life, I was transferred to the Federal Correctional Institution in Coleman, Florida.

I didn't miss a beat after moving to Coleman. And the agents picked up right where we'd left off as well, spending hours debriefing me about Chicago's criminal organization. At the time, a lot of people in law enforcement believed that thanks to Strawman, the

Outfit had gone the way of the dinosaur. But I told the agents that I knew differently, that under Hy Larner's financial leadership the Outfit had just gotten more sophisticated and more diversified, and with the FBI's help, I could prove it.

During our dozens of interviews, I walked the agents through my life inside the Chicago mob, emphasizing the fact that, if Hy Larner's operations overseas and in Panama were any indication, the Outfit had entered a new age where organized crime didn't need a crew full of leg-breakers and two-bit hustlers—not when they had big-time partners like corrupt foreign governments with entire military operations at their disposal.

The agents were intrigued, so much so that they launched a major investigation into Hy Larner and the Bastones. I could almost see those bastards behind bars. From my perspective, it was going to be a dream come true.

Over the next months at Coleman, while assisting the FBI on a number of investigations, I managed to fill several boxes with notes about my life in the Outfit. Unlike the previous years I'd spent in prison, time seemed to fly now that my life had a focus. Before I knew it it was 1998, and my release was right around the corner. I had high hopes that the information I'd provided the agents would lead to the indictment and arrest of a number of Chicago crime figures, in particular Hy Larner and the Bastones.

Just days before my release, the agents called me to a meeting. I figured this was it, the moment I'd been waiting for. But the minute I walked into that room, I knew they weren't there to give me good news. The agents got right to the point. The State Department had intervened, quashing their investigation. It seemed my old pal Mr. Larner had some very influential friends. It was over.

⊕ ⊕ ⊕

Ever since I'd agreed to cooperate with the FBI, I'd been riding high, living on the idea that someday I'd have my revenge. That's all I thought about. So I didn't take the news that the investigation had been terminated very well. It was a hell of a letdown.

I was so consumed by anger and bitterness that I didn't even care that in less than a week, on March 3, I'd be out on parole, a free man. Really, what did it matter? I'd always be a prisoner—unless I found some way to be free of my hatred for Sal and Hy. And that meant getting revenge, which seemed impossible now.

On the morning I was supposed to leave Coleman, I finally started to pack. Aside from some underwear and a pair of jeans, I didn't have much. It was all just odds and ends: some cards and letters from family and friends that I'd saved, a few pictures of Joey, the *Chicken Little* storybook I'd kept under my pillow all those years. And of course, the boxes of notes I'd made during my work with the FBI.

As I slipped the storybook into one of the boxes, I couldn't help but smile at the memory of Joey, pointing to the sky and giggling, "The sky is falling." How many years had it been since that day at La Tuna? It was funny how words stuck with you over time. And how they could bring back feelings you thought you'd left behind years ago. A few words in a book could have a special sort of power. They could make you laugh. Or cry. Or—

I stopped packing and stared down at the notes crammed inside the box. What were those old clichés? That knowledge was power, that the pen was mightier than the sword? Now that the possibility of putting my enemies behind bars had been eliminated, the only weapon I had left was my knowledge of the Outfit's operations. And couldn't that be far more deadly than any bullets or bars? Hadn't that been what the guys in the Outfit had been afraid of all along?

I had a ton of knowledge stored in my head—and in those boxes. What kind of power would that knowledge have if it was put into words? There was a new sense of purpose in my step when I walked out of Coleman federal prison later that day. I had a story to tell. And pretty soon, the whole world was going to hear it.

AFTERWORD

Anyone who believes the Chicago Outfit was finished when the convictions in the FBI's Strawman case came down is suffering from severe delusions. Like every other aspect of society, organized crime has changed with the times. The *successful* modern-day mobster looks no more like Al Capone in his chalkstripe suit and fedora than he does Tony Soprano in his polo shirt, Italian loafers, and gold chains. In fact, today's organized criminals—the good ones—don't look like gangsters at all. They look like you and me and the nice guy next door who coaches your kid's Little League team every summer.

True, Chicago continues to have its share of sleazy shysters and petty criminals, many with ties to organized crime. But law enforcement and TV shows would like for us to believe that these men *are* today's mobsters—half-witted criminals who, at their best, are a danger only to one another.

But the facts presented in *Double Deal* clearly prove otherwise. Indeed, like the wheels of justice, the wheels of organized crime in America have continued to turn. There is always young blood standing at the ready, eager to replace the old guard. The Chicago Outfit is a turnstile of constantly revolving names and faces. Slam the prison door on one man and, in the blink of an eye, another steps forward to take his place.

Such is the case with those men who have served as "boss" in

Chicago. Spanning a century, their names roll out across time, one after the other, like so many bocci balls—Colosimo, Esposito, Capone, Ricca, Nitti, Accardo, Giancana, Battaglia, Aiuppa, Ferriola. Ample proof that there is no lack of qualified prospects for the job.

In the past decade, many of the "old guard" have died. Among them:

- Turk Torello died of cancer in 1987.
- Gerry Scarpelli died in 1989 at the MCC in Chicago. Although ruled an apparent suicide, many organized crime insiders consider the circumstances surrounding his death highly suspicious.
- Charlie Nicosia died in November 1990.
- Joe Ferriola died in March 1991 while waiting for a heart transplant.
- Tony Accardo, the mob's elder statesman, died in 1992 from congestive heart failure.
- Joey Aiuppa—mob boss and sponsor of Tony Spilotro after Sam Giancana's death in 1975—died in 1992, a month after his release from federal prison.
- Pat Marcy died in 1993 after suffering a heart attack while on trial for his role in widespread political and judicial corruption, including several infamous court fixes and bribes.

Among Michael Corbitt's law enforcement colleagues, time has proved his instincts regarding their extracurricular activities disturbingly accurate, in particular his assessment of William Hanhardt. Hanhardt, the Chicago cop and chief of detectives who burglar Roger Douglas claimed had set him up, was apparently enjoying his "golden years" in retirement until he was indicted for his role as the mastermind behind one of the nation's largest organized crime jewelry-theft rings. He pleaded guilty and was sentenced in May 2002 to 12 1/2 years in prison.

Additionally, for those involved in the Dianne Masters murder, the past years have not been kind:

- Alan Masters died of a heart attack in 2000 while incarcerated at Rochester federal prison.
- James Keating remains in prison, where, according to insiders, his health continues to decline.
- Tom Scorza, the assistant U.S. attorney, was caught up in the storm of controversy that followed Corbitt's testimony in the El Rukn case and has since retreated from public life.

But what of Hy Larner? According to a retired FBI agent, the first real opportunity to expose Larner as the brilliant head of what could be the world's largest organized crime network and gambling empire came in the early 1990s, when the FBI was poised to launch a massive investigation into Larner's operations in the United States and abroad. Of particular interest to the agency were Larner's activities in Salt Lake City, Utah, where his airline pilot son, Bruce, happened to be residing. After being directed by his superiors in Chicago to forward all of his files on Hy Larner to the agency's Salt Lake City office, the agent anticipated that major inroads would soon be made in the FBI's probe of Larner's operations. Instead, the agent learned the investigation had been quashed by "someone in Washington" before it got off the ground.

This was not the first or last time Hy Larner's activities managed to escape the efforts of law enforcement. In fact, there have been several. After bestowing Hy Larner with the status of "international criminal and kingpin," the DEA launched a major investigation into his Miami and Central American operations. According to an FBI insider, this investigation was "stonewalled" by Washington, sometime after the torture-murder of Mugsy Tortorella, who, according to Michael Corbitt, was seen by members of Larner's crew in the company of DEA agents in Miami, Florida, shortly before his death.

Another, more recent and far more massive, investigation into Hy Larner's activities met a similar fate. A sixty-five-man task force was assembled through the joint efforts of the FBI, IRS, and DEA for the sole purpose of investigating Larner's Panamanian affairs. On the verge of launching the full-scale undercover operation, the task force suddenly hit a dead end when the U.S. State Department ordered the investigation terminated.

Based on these three amazing examples of good fortune, it seems safe to say that Hy Larner has been a very lucky man—or a very well-connected one. Given Michael Corbitt's observations of Larner's operations, the latter appears to be the most logical explanation. But sadly, if that is the case, then it is thanks to individuals within the United States government that Hy Larner—an organized criminal of the highest order—has remained not only a mystery man, but a free one as well.

Several key players who were involved with Hy Larner's national and international gambling operations have died, been incarcerated, or faded from view since Michael Corbitt's imprisonment and release, among them:

- Sal Bastone died at Mayo Clinic in 1998.
- Carmen Bastone oversaw the Outfit's international affairs and Hy Larner's immense gambling interests until his death from pulmonary disease in April of 2002.
- Hank Greenspun, a friend of Hy Larner, died in Las Vegas, Nevada, in 1989 at the age of eighty. In 1993, a square was dedicated in his honor at Jerusalem's Hebrew University Botanical Gardens for his contributions to the formation of the Jewish state.
- Al Schwimmer, a U.S. citizen and fervent Zionist, was convicted several times during his lifetime of violations of the Neutrality Act due to his repeated involvement in weapons smuggling to Israel. Schwimmer received a presidential pardon shortly before Bill Clinton left office in 2001.

According to FBI insiders, Larner's illicit machine and money laundering empire continues to thrive under the control of the Bastones' sons—Carmine, Joey, and Jamie—and their longtime collegue, Joey DeVita. In Arizona, Victor Vita is running a seafood company which, sources insist, serves as a front for Larner's numerous illegal activities. As for Larner's multibillion-dollar international dynasty, it is thriving as well, and according to FBI insiders, Hy's very capable son, Bruce Larner, is involved.

Although *Double Deal* sheds a new, harsh light on many aspects of organized crime's involvement with government and international affairs, as with its predecessor, *Double Cross*, we are once again left with more questions than answers. What we have learned about the man Sam Giancana once called "the Meyer Lansky of Chicago" only adds to the intrigue. The life of Hyman Larner remains a mystery. As does his "death."

In 1991, when Panamanian newspapers proclaimed the Chicago mobster "muerte," many in U.S. law enforcement believed this was merely a ploy, that in an effort to elude authorities, the crafty mobster had staged his own death. And indeed three years later that theory seemed to be confirmed when rumors surfaced that the elderly mafioso had been discovered alive and well, residing in—of all places—Flathead, Montana. So which was it? Was the man who'd been designated an international "kingpin" by the DEA alive, or dead? Given Hy Larner's past, most insiders were keeping their money in their pockets.

Then, in the fall of 2002, with *Double Deal* on the verge of going to press, everything changed—including this book's ending— when Hy Larner's name abruptly appeared in the *Miami Herald*, listed among the obituaries. It was just a few lines. Mr. Larner, the paper stated, had passed away on October 12, 2002, and would be laid to rest at Memorial Park Cemetery in Skokie, Illinois. Strangely, there were no front-page headlines heralding the death of one of the nation's most powerful mobsters. There was no fanfare. Or garish floral arrangements. No press, no cameras, and no microphones—none of the media circus that has become so typical of a prominent gangster's funeral—just a small circle of family and friends who came to pay their respects.

It seemed a fitting end to a life spent avoiding the glare of publicity. But was it really the end? For a man like Hyman Larner, who excelled in sleight of hand, anything seems possible. After all, this was a man who had orchestrated his own demise once before, in Panama—so why not now, just as his darkest secrets were about to be exposed to the entire world? Surely there would have been another investigation into his operations. Perhaps this time he wouldn't have been so "lucky." Perhaps this time Hy Larner would have found himself behind bars. And as any wiseguy will tell you, the joint's not kind to the old-timers; they die inside. That seems motive enough for a man who has never known the true value of freedom.

In the end, whether Hy Larner is dead or alive is irrelevant. Whether he's lying in a coffin somewhere in Skokie, Illinois, or on a sunny beach in Panama doesn't really matter. What does matter, however, is that the illicit alliance he and his cronies forged some fifty years ago with international leaders and rogue elements within U.S. Intelligence and the military *is* alive.

The fact that Hy Larner and his global network of gambling operations have remained virtually unknown and unimpeded by the law for almost half a century is a shocking testament to the tremendous failing of our nation's leaders as well as its law-enforcement agencies. According to mob and FBI insiders, the Bastone crew remains involved in all illegal gambling activities west of the Mississippi, including Las Vegas, as well as internationally and Bruce Larner, Hy's son, has been associated with these activities as well. So the big mystery isn't who is (or was) Hy Larner, but why his operations continue to thrive. Federal agencies know the perpetrators' names. They know where they live. And still, they do nothing.

It is our sincere hope that, with the publication of *Double Deal*, the pursuit of justice in this matter—having been so clearly abandoned by those individuals whom the American people have entrusted for decades—might be taken up once more.

ACKNOWLEDGMENTS

Michael Corbitt:

People who helped knowingly and unknowingly, J.B., J.O., F.M., J.K., R.W., J.B.-1, K.M., L.T., D.P.

Stuart Kaminsky, a great author, and his wife, Enid, who kept the project alive and opened their hearts to me. They were also frequent visitors while I was away, and after I was released they invited me into their home.

All of my attorneys who worked to bring me home; some for no money, most for no money.

Dennis Berkson, who is a genius and a good friend.

William Biederman, who fought the good fight with NBC-TV. John Ehrlich, also a genius and a great attorney, and now my friend. Nan Nolan, a great attorney and now a federal judge in Chicago. Amy Y., an attorney who came into my life like a tornado and was instrumental in getting me through all the legalities of doing a book deal. And taking me grocery shopping for the first time in eleven years.

Father and son Lou and Nick, who came many miles many times to visit me in many ungodly places: true friends, not users like most of the people in my life.

My sister Ceil, who was always there no matter what.

M.C., who I vented to for years and who took me for me and never judged. Fred and Elaine, who came thousands of miles to see that I was eating well.

J.P., who helped me in too many ways to count.

And special thanks to:

My agent, Frank Weimann, who flew from prison to prison—and is a bad flier—to keep this project alive and then shopped the book till he found someone courageous enough to publish it. As you can see, he was successful.

Mauro DiPreta of HarperCollins for having the Courage.

Joelle Yudin, Debbie Stier, and Suzanne Balaban for all their hard work and effort on the book.

Sam and Bettina, with whom I spent so many hours reliving so many things I would like to forget and through that became part of me and will be forever.

Uncle Lester, who gave Joey his time and me many laughs when nothing was funny.

Sam Giancana:

In 1992, after the publication of *Double Cross,* the story of my uncle and godfather, Chicago mob boss Sam Giancana, I was left with more than a few burning questions. For one thing, I wanted to know who killed Sam Giancana. But even more than that, I wanted to know what happened *after* his grisly murder in 1975. Did organized crime and its ties to the CIA and world affairs die with him—or did the Chicago mob continue to thrive, growing more powerful, and more deadly? And if so—how? And under whose command?

Unfortunately, my initial attempts at uncovering answers to these questions turned up little of substance and, after several years, I was discouraged and frustrated, particularly since I'd hoped my investigation would lead to a suitable follow-up to *Double Cross.*

I'd almost given up on the entire idea when my agent and friend Frank Weimann of the Literary Group introduced me to Michael Corbitt, a mobbed-up cop out of Chicago who had a hell of a life story—as well as some of those answers I'd been looking for. For this introduction, I will always be indebted to Frank.

Acknowledgments

I am also deeply grateful to William Morrow and its visionary publisher, Michael Morrison, and executive editor Mauro DiPreta. Mauro's talented editing, untiring curiosity, and drive for perfection have, without question, been largely responsible for making *Double Deal* the gripping page-turner we'd all hoped for when this project first got under way. Mauro's generous spirit and courage have won both my respect and heartfelt gratitude. Morrow's assistant editor, Joelle Yudin, has also engendered my unflagging admiration. Joelle's unbridled enthusiasm for this book, coupled with her professionalism, has made working with the publisher a true pleasure.

I also wish to thank the countless FBI agents, undercover resources, and numerous journalists—most of whom cannot be identified—for their assistance in this project. Early on, these individuals recognized the importance of this undertaking and provided the factual and historical information necessary to back up Michael Corbitt's many anecdotes. Among those who can be mentioned are retired FBI agents Jack Bonino, Jack O'Rourke, and Frank Marrocco and the noted author and reporter for the *Chicago Tribune* Ray Gibson. I also want to express thanks to the Chicago Crime Commission, the Chicago Historical Society, the *Chicago Tribune,* and the Chicago *Sun-Times* for their cooperation and assistance.

And last, but certainly not least, I must acknowledge my deep appreciation to Michael Corbitt, whose wild ride through a life of double deals has made for one of the most amazing books you'll ever read.